Kids & Teachers
TARDIGRADE
Science Project Book

HOW TO FIND TARDIGRADES AND OBSERVE
THEM THROUGH A MICROSCOPE

Michael W. Shaw

Fresh Squeezed Publishing
Richmond, Virginia

Michael Shaw/Fresh Squeezed Publishing
P O Box 742
Midlothian, VA 23113 USA
www.mikeshawtoday.com

Ordering Information:
Quantity sales. Special discounts are available on quantity purchases by corpora- tions, associations, and others. For details, contact the "Special Sales Depart- ment" at the address above.

Kids & Teachers TARDIGRADE Science Project Book/ Michael W. Shaw.
1st ed.
ISBN-13: 978-1499134988

Contents

To Erin Lee

Education is not the filling of a pail, but the lighting of a fire.

—WILLIAM BUTLER YEATS

CHAPTER 1

About Tardigrades

Before we can do a classroom science project or enter a tardigrade experiment in a science fair, we need to know a few things first. We need to know how and where to find tardigrades, how to collect and view them, how to preserve them, and how to take pictures of them. So let's start with these basics.

The word "tardigrade" comes from "tardi," meaning slow, and "grada," meaning to walk or step. Tardigrades have also been called "Moss Piglets," because they are found in moss and have a snout that kind of looks like a pig's nose. I don't think they walk so slowly, but when compared to other micro-life (which speeds around the microscope slide), yes, tardigrades do move more slowly.

What used to be mostly an unknown species is quickly becoming a popular animal and household word: tardigrade. Google it and you'll see how much information is out there. My website (listed at the end of this book) tries to keep up with all the cool things about tardigrades, such as a song about them, a knitted tardigrade sweater, articles and blog posts, toys, television shows, and many videos on YouTube.

Tardigrades are microscopic creatures that are a maximum of one millimeter in size, but usually are found to be about half that size. They seem to be related to either nematodes (for example: roundworms) or arthropods (such as crabs, water fleas, ostracods, or insects like ticks or mites).

Here on the right is an ostracod (or seed shrimp). Below, you can see a water flea (or daphnia) with eggs inside. Are these creatures related to the tardigrade?

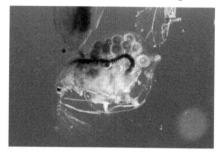

Ostracod and Water Flea. Below is a tick. Any of these related to a tardigrade?

The picture on the right is a type of tick is called a "dog tick."

Look below at the side of the theory that tardigrades may be related to nematodes. I've had to piece together three images of a nematode, because this creature is so very long under the microscope.

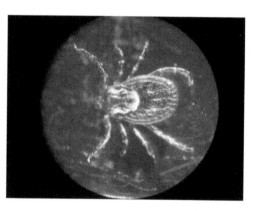

Nematodes are also known as roundworms, and there are over 25,000 species. Half of these are parasites, not necessarily human parasites, though. Because of some similarities between tardigrades and nematodes, it is thought that tardigrades may have evolved from nematodes. Tardigrades are harmless, however, and if you

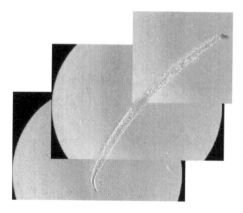

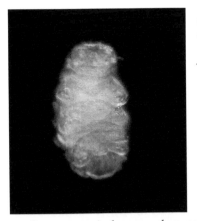

swallow one, you don't need to worry. They are not parasites.

Why Are Tardigrades So Interesting?

One reason is because they are similar to us in some ways. They have a mouth, an alimentary tract (that's the throat to the stomach tube), and they digest food and excrete it like we do. They have a nervous system and eyes (though primitive). What really

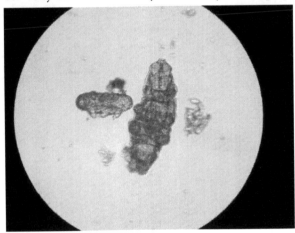

makes tardigrades fascinating, however, is their cuteness!

At the top of this page is a baby tardigrade on his back. Don't you just want to tickle him?

They are often referred to as "Water Bears" because they look like chubby little bears with eight puffy legs, and they have claws that look like those a grizzly bear would have. With their two eyes, brown color, and hungry mouth, one is reminded of a bear. They even move around like bears.

Don't get between this mother bear and her baby (picture above), or you might be attacked. That was a joke. Tardigrades do not attack.

What Do They Eat?

They eat other micro-life, such as rotifers and nematodes, which you will

almost always find where you find tardigrades. Tardigrades have also been reported to eat other tardigrades. I have observed green matter in their digestive tracts, so they probably eat vegetation as well.

Just above is a picture of a rotifer. Possibly this might be a meal for a tardigrade.

How Do They Grow?

They molt their outer skin. Here is a picture of a tardigrade about to shed its old skin. Look at the tardigrade inside its own husk, or "cuticle" as it is called.

How Long Do They Live?

Instances have been documented where tardigrades have lived for a hundred years. But we have to be careful when we say that because surviving for a century is not the same as an active lifespan. A turtle or a parrot, for example, can live as long as a hundred years and, as a pet, can outlive its caretaker. We imagine this type of pet as moving around and living a pleasant life of pampering, attention, and eating good food. On the other hand, a tardigrade will go into a type of hibernation called "cryptobiosis" sometimes for decades. So the lifespan of a tardigrade may actually be much shorter than one hundred years if the tardigrade were kept moving around in a petri dish as a pet.

How Do They Reproduce?

They lay eggs. In fact, it is sometimes so difficult to identify types of tardigrades that you have to rely upon their odd shaped eggs. Many of their eggs have points like spikes. The below photo shows two examples, taken with transmitted light called "brightfield lighting." Light shines under the specimen and passes up through the egg making it transparent so you can see the spikes.

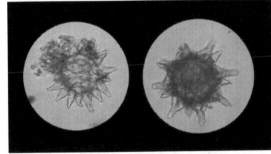

There are other types of eggs with a round smooth exterior like those in the photo of the egg cluster at the top of this page, taken with incident light called "darkfield lighting," because the background is dark and the light comes from the side or top. You can't see through the eggs, and this type of lighting shows you their texture.

Where Do Tardigrades Live?

It seems that tardigrades live everywhere, and have even been found in 90 million year old amber. Like sharks, tardigrades have a prehistoric ancestry.

Below is a picture of an ant I found in a 90 million year old piece of amber. I was looking for tardigrades, of course, but I did not find any. This amber tells me, though, that ants are also a prehistoric species.

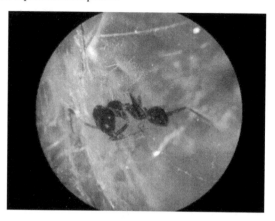

Tardigrades have been found in Antarctica, all over Europe and Asia, and throughout the USA. My research involved a survey of New Jersey, visiting all 21 counties, collecting specimens and checking for the presence of tardigrades. Guess what? I found them in all 21 counties. Therefore, if you are looking for tardigrades (because you're reading this book), you will surely find them in most forests and other natural outdoor places you search.

How Did They Get to So Many Places?

It's easy to imagine. A tardigrade, when in the "tun" stage, is tucked into a tiny ball, like the smallest particle of dust. The photo above shows how a tun looks.

Dust (which can contain pollen, ash, dust mites, sand, animal hairs, pieces of dried skin, pollutants, and tiny tardigrade tuns) travels thousands of miles. This

is how we believe tardigrades spread across the earth. Since even volcanic islands yield tardigrades, we are fairly certain that the winds, and perhaps birds, have spread them. Tardigrades are certainly not an endangered species.

A tun is as small as a grain of pollen, so it can float on air like a bubble in the wind. We know that pollen is spread by birds, insects, and by the wind. This bird is sitting on a branch that is encrusted with lichen. Since lichen is a likely place you will find tardigrades, there is a good chance that bird claws could carry tardigrades or their eggs. A bird may even fly somewhere to build a nest using soft lichen, and thus carry tardigrades to another location. To the right is a picture of sunflower pollen.

Look at all the spikes. These pollen grains look almost like tardigrade eggs, right?

Tardigrades Withstand Extremes

Tardigrades can withstand heat of up to 120 C (248 F) and cold down to minus 272 C (-457 F), so you should have no trouble finding one in Summer or Winter.

Besides the extreme temperatures, tardigrades have been tested with X-Rays, harmful ultraviolet light, and deadly radiation. They have survived high doses of each. They can also survive high vacuums (like that in space) and great pressure (like that under the deepest oceans). How do they do it?

They protect themselves by going into a type of hibernation called "cryptobiosis." That is when they roll up into a dried little ball (a tun) and just stay dormant, with no sign of life whatsoever. You have seen a picture of a tun in a previous photo. Scientists are said to have rehydrated tardigrades from a piece of moss in a museum collection that was a hundred years old.

Since tardigrades can withstand so many extremes, you will find them outdoors almost any time you look for them, everywhere and in every season.

Did They Come from Outer Space?

Because of these abilities to withstand so many extremes, and most importantly the ability to survive for a hundred years, theories about tardigrades coming to Earth from space have recently been proposed. The following explanation will help.

The closest star to our Sun is about 4.2 light years away. A light year is simply the distance light travels in a year (moving at 186,000 miles per second or 299,338 km/second). A light year is a measurement of distance (not time), like a mile or a kilometer.

How far is 4.2 light years? About six trillion miles (9.4 trillion km). Without getting too bogged down in a math and technology discussion, it would take our fastest spaceship about 55,000 years to make the trip.

Can a tardigrade survive that long? In the extreme cold and vacuum of

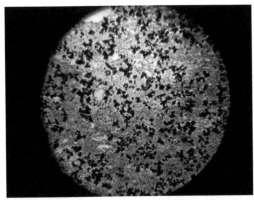

space travel, possibly. But what about the heat of a meteor hurtling through Earth's atmosphere? Could a tardigrade ride on the surface of a meteor? The surface of a meteor is about 3000 F (1650 C) when it burns through Earth's atmosphere. It is so hot when it falls through the atmosphere, not even a tardigrade can survive that heat. A meteor gets so hot that the heat will burn off anything clinging to the surface. Some say the inside of a meteorite remains cold from outer space. If so, then some organism or DNA might be preserved in that long journey. We have only a few reports by people who have touched a meteorite soon after its landing, so we don't know for sure what the internal temperature is. But we have looked closely at the insides of meteorites.

Have you ever seen the inside of a meteorite? The pictures on this page show you how they look inside. Meteorites are mostly iron and nickel, or made of rock. We cut them up into very thin slices and look at them under the microscope to see inside. We have yet to see anything that looks like a living organism or a fossil of a living thing.

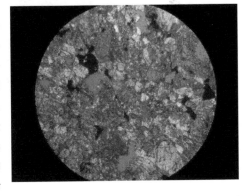

Here are some pictures I've taken of the insides of meteorites. The meteorite in the picture at the top of this page fell in the Chicago, Illinois area. The meteorite in the picture at the bottom of this page fell in Northwest Africa. Tardigrades may be everywhere, but they have not been found inside meteorites. The insides of meteorites do not look like slices of amber with preserved animal life.

Well, what's the answer? Did tardigrades come from space? Most assured-ly, they did not. Besides the above reasons, we already know that tardigrades are closely related to species on Earth. We don't know exactly how and precisely to which species they are related, but DNA results are telling us more and more about how tardigrades are related to other creatures on Earth.

There is a theory that all life on Earth originated in outer space. That theory, called "Panspermia," presents the possibility that some very low life form, like bacteria or even DNA fragments from another planet, arrived on Earth and all life evolved from that. Well, tardigrades would be included in that evolution, so only in this case can you say yes, tardigrades, *with everything else*, came from space. Google "First Animal to Survive in Space" to see a video about this theory.

Now let's move on to finding some tardigrades right here on Earth.

CHAPTER 2

Where to Find Tardigrades

If you are entering a science fair, you may be tempted to order tardigrades from a biological supply company. You can do that if you need a lot of tardigrades all at once for your project, or if the project requires all of the tardigrades to be of the same species.

There is no need, however, to add extra expense to your science project when you can find tardigrades easily everywhere. Yes, tardigrades are everywhere (to be scientific, that means everywhere we have looked). If you go to a

nearby park or forest, you will most likely find tardigrades. If you have trees in your front yard, you really may not have to go farther to find tardigrades.

Trees

Take a look at this tree. It is the perfect tree for tardigrades. Why? To find tardigrades, you will have to find the correct type of tree- one without smooth bark. Trees with rough surfaces have two things that support the presence of tardigrades: moss and lichen. Although tardigrades may be found on the smooth surface of a Sycamore tree, like the one pictured below, it is not likely. A few may be on the surface of this white Sycamore tree, but it will take

too much effort for you to find one. Instead, look for trees which have bark with deep furrows, and lots of texture, as on the previous page.

Below you see a tree with some juicy moss sur-

rounding hard crusty lichen. Moss samples from any tree are fine. Here, the type of tree surface doesn't matter be-cause the moss will pull off the tree easily without a lot of dirt, and moss is a good place to find tardigrades. You might also find a tree that has some moist lichen grow-ing on it, that greenish yellow

fuzzy stuff which you see on the shady side of a tree, usually growing near the bottom (pictured on next page). The bottom of a tree retains most of the rain run-off and moisture, so lichen tends to grow at the lower portion of the tree.

Lichen

What is lichen? Technically it is a combination of algae and fungus living together in what's called a "symbiotic" relationship. Lichen, just like the tardigrades which we find living in it, grows in the most extreme places on Earth. Lichen thrives in the desert, in rain forests, the arctic, on rocks along the sea coast, in woodlands, and even on toxic waste dumps. You will find lichen growing on trees (pictured on right), on stone and brick walls, forest rocks, on telephone poles, and on boulders in the

town square. Check out this lichened brick wall which is probably loaded with

tardigrades. Once you realize which is the best type of lichen, you will always be on the look-out for it. You might carry a little specimen collecting kit in the car, with a small knife, some envelopes, and a fine point marker, just in case you see a good lichen sample somewhere.

As you can see from the picture below, some lichen is hard and crusty. No tardigrades here I would bet. Instead, look for the fluffy kind of lichen, which after a rain will stay moist. That's where tardigrades prefer to live. I've looked in many samples of the dry type of hard lichen, and found very few tardigrades. Yet, I have always found lots of

tardigrades in fluffy lichen. The best lichen is greenish yellow to bright yellow, the softer the better. A good time to collect lichen is the day after a rain. It will still be moist, and chances are you can find tardigrades in just a few hours.

Leaf Litter

Water bears like moist places to crawl around in, so keep in mind that they also live in moist leaf litter besides puddles and streams in the forest. They live in the gutters along your roof too. If you are collecting leaf litter in drainage areas, you do not need a great deal of material. Collecting leaf litter is a bit messy, and if it is wet, you will find it challenging to transport. Although you are guar-

anteed to find many interesting forms of micro-life in leaf litter, you may find few tardigrades.

Moss

When collecting moss, look for soft young moss. Once you look at a bed of moss, you will actually see a mix of new growth and some older moss that tends to be dried out at the top because older moss is taller. The below picture shows

some moss growing on bark. In this case, you would select the moss sprouts at the edges because they will be moist. Since a sample of moss usually comes with a lot of

dirt, try not to pick up too much dirt when collecting the moss. The easiest moss to collect is that which grows on trees or rocks, as in the previous photo. There is really very little dirt to contend with and you would just get mostly the green material. Remember to look on stationary rocks, the rocks which do not

roll, because a rolling stone gathers no moss.

Take a look at the difference between a moss sample and a lichen sample in the water in a petri dish (picture above). The lichen sample is in the dish on the left, and the moss sample is in the dish on the right.

You can see how much more difficult it is going to be to search through all of the dark brown matter in the petri dish containing moss. Remember that under the microscope, everything is gigantic and those little grains of dirt are giant boulders that get in the way of finding tardigrades.

Ticks

When collecting any samples from woodsy areas outdoors, be absolutely sure to check yourself and others for ticks. Ticks carry several diseases, especially Lyme disease, and you really need to be very observant by checking your legs and socks. Check during and after your expedition. Wear light colored pants when going into the woods, because you can spot ticks more easily on your pants legs. Use insect repellant, and spray some on your lower pants legs as well. Lyme disease is transmitted mainly by deer ticks, wherever deer may roam. Incidents of Lyme disease have been reported all over the USA. This is not to scare you, but it's a safety concern.

Besides wearing light colored clothing, if possible, tuck the legs of your pants inside your socks when you walk through the woods. Before you get in the car, and once home, check yourself again. I once had a tick crawl up my pants and hide at my waistline. To the right is a picture of a deer tick. (USDA Photo by Scott Bauer.)

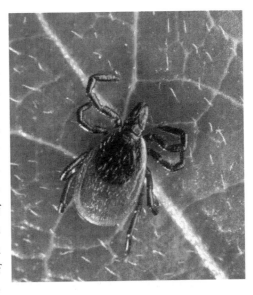

The symptoms of Lyme disease do not always appear immediately, sometimes appearing a week later, sometimes not appearing at all. If you are bitten by a tick and you remove it, try to save the tick in a carefully wrapped piece of paper, or in one of the specimen envelopes that you are using. Later, you can identify the tick using a magnifying glass and that could help you or your doctor know how likely it is that you might have Lyme disease.

This deer probably has ticks carrying Lyme disease. A walk through these woods could leave ticks clinging to your pants, so be careful.

Now that you know where to find tardigrades for your science project, the next chapter will give you some techniques to collect them.

Collecting the Specimens

Y ou will need something to collect the specimens in, something to write with, some sort of scraper like a knife or razor blade, and a way to know your location – either a GPS device, smartphone, or...a map.

Bags and Envelopes

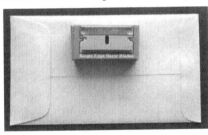

You can use school lunch bags, the inexpensive paper ones, or small envelopes whether you are collecting lichen, moss, or even bark scrapings. Don't use plastic bags because the moisture that remains in the plastic will allow mold to grow. You should use school lunch bags or little coin envelopes. Coin envelopes are small, about 3 by 5 inches; they hold plenty of material, you can write on them, and you can store them easily.

Scrapers

You will need to use a sharp knife or fresh single edged razor blade. Of course, you must be super careful with either a knife or razor, and younger researchers need to get adult supervision for this step. Collecting tardigrades is a good way to introduce care and safe handling of a pocket knife if you are a parent or scout leader (assuming an appropriate level of maturity in the student).

If you are simply collecting tardigrades just to have a look at these fascinating creatures, then a pocket knife will be fine. If, however, you are doing a science fair project or scientific research, you must have a clean blade for each sample you take. Otherwise, you could contaminate one sample with another, and you would not really know where you found your tardigrades. Handle this challenge in several ways: You could use a fresh new disposable razor blade for each sample, or you could use hand sanitizer and clean tissues (or packaged alcohol wipes), to carefully clean the knife blade between samples.

Navigation

It's a great idea to carry a small handheld GPS device with you any time you go into the woods. I've done quite a bit of hiking, and as much as I like my smartphone (which has cool hiking apps), phone batteries do not last very long, especially when using navigation programs. A GPS device will give your loca-

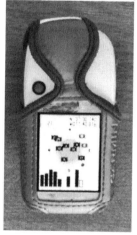

tion as well as provide the exact GPS coordinates of your tardigrade specimen. Simply carry extra batteries for the GPS device.

A mobile phone may work just fine if you are only collecting one or two samples. Smart phones will provide GPS location if you use the right app. But for a long hike, I always hike and explore with my GPS unit, and a compass.

Sometimes, even a GPS unit will fail, and it is therefore important to have a compass whenever you are hiking.

I remember one time hiking the along a small part of the Appalachian Trail, right next to a big campground. How could I get lost? I even had a trail map with me. I took a few steps into the woods because I wanted to take pictures of some flowers in a clearing. The distance was perhaps twenty steps. After a few minutes of photography, I was lost. Seriously, I had no idea where the trail was. This is because once you leave the trail, you can no longer see it from any other place in the woods. I realized then how important it is to check your compass before you leave a trail, and then you can see exactly which way to go back to the trail.

Fortunately, I had my GPS. I just looked at my track and followed it backwards to the trail. Lesson learned, and now shared with you. Always carry a compass when hiking.

Maps

If you do not have a GPS device or a smartphone, you can use a map. You can obtain a road map, you can purchase a topographical map in a hiking and backpacking store, buy an aeronautical map at a small airport, or get a free map from the visitor's center in your state. You can easily determine your own location as you collect each specimen and make a mark on your map. Later, when you get back to your computer, you will be able to look up online the exact GPS coordinates of where you were. For most purposes this is a perfectly fine way to do research. Young researchers should get used to using maps and reading the legends on a map.

Writing It Down

You will need a pen or marker to make notes and write down data on your collection envelope. Pens and markers dry out, and sometimes pens will leak. They are great to use when fresh, however if you leave a pen in a hot car, you can have problems. That's why I always have a pencil as a back-up. The faithful pencil always writes, and even if you don't have a pencil sharpener – you can sharpen it with your knife or blade. If you keep a tardigrade specimen collecting kit in the car, then keep a pencil in there as well. Once you collect your specimen, you will need to write quite a bit on your bag or envelope. For a science project, you must document each sample collected. Seal your bag or envelope with tape. First, you will give your specimen a number, and write that on the outside of the bag or envelope. This information will help you later if you decide to make a permanent microscope slide, or if you take pictures or want to save the images on your computer. That specimen always has the same

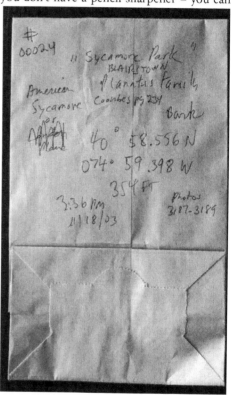

number whether it is in a photo or referred to in notes. Next, write the date, time, and where you collected your specimen. It doesn't have to be super neat, but you must have all the important data there. You can use a GPS device to write down the coordinates, or as mentioned earlier, you can mark the location on a map (using your specimen number just assigned). After you mark your good old fashioned map, you can look up the coordinates online, and then you would go back and write those coordinates on the bag or envelope. Using a map is not as accurate as GPS, but it will save you money because you can usually get a map for free, and the GPS coordinates online are free. Look at the paper lunch bag on the previous page, which I used for collection: It is marked with place, tree type, sample type, specimen number, date, time, latitude, longitude, altitude (those last 3 items from GPS), and photo numbers from the digital camera which was used to take pictures of the tree.

Collecting Moss

If you are collecting moss from the ground, no tools are needed. Simply pull out some clumps of moss with your fingers. Just a little bit of moss will be fine,

and it's best to avoid big clumps of dirt.

Remember that tardigrades cling to the green moss leaves, so the dirt is unnecessary.

You'll end up with lots of dirt anyway, but do try to avoid it. The idea is to collect the smallest sample of matter that will yield the most tardigrades, and perhaps even some tardigrade eggs. If you are collecting moss from slippery rocks in a stream you will have to be very careful. While it may look tempting, slippery rocks are dangerous and it's not worth the risk to get a small sample of moss.

Collecting Leaf Litter

Collecting leaf litter is messy because when you have leaves that are wet, they are difficult to transport. They will also collect mold quickly and this will actually kill the tardigrades. You must be careful too. Collecting leaf litter from

a roof can be dangerous, and even when you collect leaves on the ground there are hidden dangers like snakes.

Collecting Lichen

To scrape lichen off a tree or brick surface, start by holding the blade at a 90 degree angle to the lichen. You don't need to slice the lichen off the tree; instead use the edge of the blade to scrape it into your bag or envelope. The motion is like a sweeping motion, where the sharp edge of the blade is like the bris- tles of a broom. See the above pic-

ture. Sometimes you find an opportunity to collect tardigrades from moss on the ground and from lichen on a tree at the same time. See the photo below. It's fun to compare and see how many and what kinds of tardigrades you will find in each specimen collected from the same locale. You get to compare habitats and learn the skills you will need to collect and ultimately rehydrate tardigrades from moss and lichen.

As you collect the different types of specimens you become more and more efficient in the process. You will want to collect specimens wherever you travel, whether it is on a family trip or vacation, or on a business trip, or even at a friend's house. Remember to be respectful of other people's property, and ask permission before collecting specimens on private property. For my research, I relied on public parks, city streets, national forests, parking lots, and other plac- es that were open to the public. There are plenty of places to scrape lichen and pluck moss without trespassing or violating someone's private property.

CHAPTER 4

Rehydrating the Tardigrades

Now that you have collected some specimens in envelopes or paper bags, it is time to find out if you have any tardigrades in those specimens.

You will need a clear plastic petri dish or something like it. You could use the cut off bottom of a plastic cup, the top of a little plastic cosmetics case, or even a piece cut out from blister packaging. Most small products we purchase are "blister packed." This is packaging with a hard plastic shell which holds the product, backed by a colored cardboard product description. Cut out a piece of plastic from the blister pack that will hold about a half inch of water, like a little bowl. The "dish," as we will call it, does not have to be round.

Here are some pictures of various options for your dish. A trip to an art supply store or crafts store will definitely give you something to work with. Look in the bead section of the store. Here you see one such container that has the perfect lid.

For a science fair project, I recommend that you buy plastic petri dishes (pictured below) from a science catalog or online. I have some available in my Amazon store (listed in the back of this book) as well. These are very inexpensive, and you get at least a dozen in a package. It looks very professional in your presentation.

If you like free, you can cut off the bottoms of clear plastic drinking cups, which would work too. Finish your drink first, okay?

Most fast food places have little condiment cups like the one shown below. These are also excellent for tardigrade research, and I know of one university scientist who uses them all the time. These condiment cups are good for starting a suspension, however they are a little too deep to use when actually searching for tardigrades. They are easy to

write on and good if you are doing three or four suspensions at the same time. They are not, however, good for observing tardigrades because the sides are frosted and the bottom is not flat. You will have to pour the liquid and particles into a flat dish, probably a petri dish or blister pack section in order to actually see tardigrades.

Once you have decided upon your receptacle, sprinkle some lichen or moss into it, just a tiny amount. Remember, this is a microscopic journey, so a pinch is a great deal of material under the microscope.

This part does get a bit messy, so plan on doing this step on a surface you can wipe clean when you are finished. Dried lichen and moss has a tendency to carry a static electrical charge, and jump to the lid and sides of your petri dish, as well as to other surfaces as soon as you allow it to play outside of the envelope. In the picture here, you can see how the lichen particles scatter all over.

To the lichen or moss in the petri dish, add some bottled water, just about a quarter inch to half an inch. Many scientists use distilled water, however for our purposes, ordinary bottled spring water is fine. Tardigrades live outside, and their water is not distilled. Spring water is the closest thing to water in their natural environment. It has natural minerals like you find in water outdoors.

At this point, you will notice that many of the moss or lichen particles tend to float on top of the water. Water has something called surface tension, and this is acts like an imaginary film that prevents really small particles from sinking. If you visit a pond, you will see certain insects scooting along the surface of the water, not sinking. They are riding on the surface tension of the water. The most well-known of these is the water strider. These insects have very long thin legs that allow them to balance on top of the water's surface without sinking. It is the surface tension of the water that holds them up.

After a while, all of your specimen particles will absorb some water and sink. If you are in a hurry, you can fill a small sprayer bottle with water, and spray the floating lichen or moss. It will sink a bit faster because the water spray will break the surface ten-

sion and soak the dry particles that are floating.

Now, having added the water to your specimen dish, let it sit. This is called a "suspension." Set the dish aside and let the lichen or moss soak overnight. Another reason to consider using plastic petri dishes is because you can cover these to prevent dust and mold spores from getting into the water. If you are using an open cup or lid, however, it's not a major problem because it takes mold a few days to grow, and by then you will have found your tardigrades. Does overnight seem like a really long time to wait? It is a long time, and you might be impatient. There is a chance that you will not have to wait, and you might find tardigrades in as little two hours, so there is no harm in checking your specimen dish every few hours as described in the next chapter. If your lichen or moss sample is fresh, taken perhaps after a rain the previous day, you might be lucky. Generally, though, you are assured of finding tardi-grades by letting your specimen soak in the dish overnight.

Next we move on to learning how to find tardigrades in the suspension.

CHAPTER 5

Your Microscope
and Accessories

Y ou will need a microscope. If you are a teacher, you can use whatever your school has, and the magnification will be just fine. You do not need higher powers to see tardigrades. A "dissecting microscope" is

best for spotting tardigrades, and there are some very good amateur dissecting microscopes available at reasonable prices. Sometimes called a "binocular stereo scope," a dissecting scope will open up a whole new world to you. This type of microscope, shown to the left, has a lot of space between the lens and the specimen platform. The specimen platform is close to the table, and the dissecting scope can magnify all sorts of things beautifully in 3-D. It is good for searching for tardigrades, because you will be placing a petri dish containing water on the stage, and you do not want this to get anywhere near the lens. The dissecting scope is made especially for the purpose of magnification at a long working distance from your specimen. Aside from tardigrade projects, you can use a dissecting microscope for many types of science projects. You could do a project on seeds, or cereal grains as shown in  this picture. Look at the distance from the grain filled petri dish to lens of the dissecting scope. You can even use it around the house; it is fantastic for looking at jewelry that might need cleaning or a repair, or for inspecting a fingernail or a cut, or for looking at a splinter, even for inspecting small parts, tools, or electronics. If you find a tick on your clothing, you will be able to identify whether it is a Lone Star tick, a deer tick that may be carrying Lyme disease, or a dog tick. If you have to repair eyeglasses or a small piece of jewelry, the dissecting microscope makes it easy. Once you have one of these microscopes, you will want to put under it every insect you find, every seed, leaf, 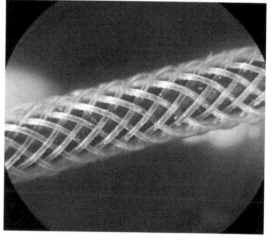 hair and fiber, and every strange piece of lint. Take a look at this piece of fly fishing line under a dissecting microscope. It is truly fascinating.

Finding tardigrades is best done at 20 to 30 times magnification. Most dissecting scopes are within that range. Pictured here is a professional dissecting

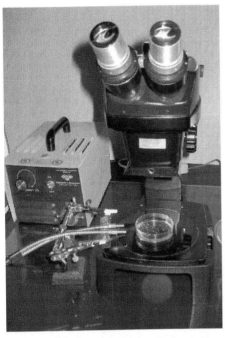

microscope. On the base, you see a petri dish with a tardigrade suspension. To the left is a fiber optic cable to illuminate the petri dish. While this is a very high quality microscope, you don't need anything so fancy. You do not need fiber optics either. In the picture below, my daughter used a less expensive dissecting scope for her science fair project on soil types in the area. This is an inexpensive dissecting microscope. Here, you can see she is using a plastic condiment cup, illuminated from the side with a fiber optic cable. You can also use a small LED flashlight

as well. In the picture, you can also see my regular "compound microscope" on the left edge of the photo. This compound scope is the kind we most often see. You will see a diagram on the next page. A compound microscope has an eyepiece at the top of the tube, and interchangeable lenses at the bottom, like the lenses on a camera. These lenses swing around on a turret to change the magnification. In the picture on the facing page to the right you can see how the entire compound microscope looks. Also, you can

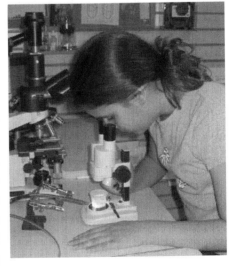

see the lens turret. You can also use a compound microscope to observe tardigrades instead of using a dissecting microscope.

The Compound Microscope

How is the compound microscope different from a dissecting scope? You can tell by the photos. You should, however, be familiar with the parts so you can better understand the advantages of each type of microscope as we move along through this chapter. On top is the eyepiece. This is the same in both the compound and in the dissecting microscope. These eyepieces are usually 10x (or 10 times) magnification. A compound microscope has one or two of these eyepieces on top (one for each eye), and sometimes three so that an eyepiece can be used for a camera (as on the facing page). When you look through the two eyepieces of a compound scope, you see the image, however it is not in 3-D. It's like looking at a flat photo. But a dissecting scope has two eyepieces *and two matching objective lenses* that allow you to see in 3-D. It is just like going to a movie theater and watching a movie on the big screen in 3-D (when you wear the special 3-D glasses).

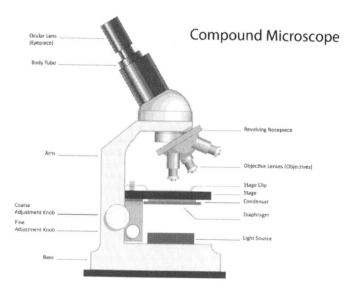

Compound Microscope

Ocular Lens (Eyepiece)

Body Tube

Revolving Nosepiece

Arm

Objective Lenses (Objectives)

Stage Clip
Stage
Condenser
Coarse Adjustment Knob
Diaphragm
Fine Adjustment Knob

Light Source

Base

The objective lenses on a compound microscope (in middle section of the scope) are each of a different magnification. These rotate on a turret, and that is how you switch them into place. It is also called a revolving nosepiece. The dissecting microscope does not have this revolving turret because the objective lenses on a dissecting scope are fixed in place, just like on a pair of binoculars. Binoculars have two eyepieces and two matched big lenses on the front that do not move. Some dissecting scopes allow you to zoom in to increase power. Both types of mi-

croscope have a stage for placing the specimen. A compound microscope has a very short distance between the stage and the objective lenses. If you are using a compound microscope like this one just above, there are some cautions you need to know about because the stage is so very close to the lens turret. This means you have a small working distance. If you put a petri dish on the stage, you might accidentally let one of the lenses come in contact with the water, and that would ruin the lens. Also, you could accidentally spill the dish of water on the stage. This could soak into the stage and rust it, or ruin the lighting assembly or electronics. Take a look at the objective lenses in the picture on the next page. These lenses come in very close contact with the specimen. That is okay when the specimen is flat and on a microscope slide which is very thin. If, however, you are looking for tardigrades in a petri dish full of water you have to be

very careful not to let the objective lens touch the surface of the water. Those longer lenses sticking out at an angle near the stage are also simply called "objectives." The higher the magnification, the longer 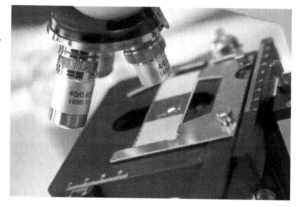 the objective lens, and the closer it must be to the specimen. If you want to look at a petri dish of water, you must use the shortest lens barrel. In the picture directly above, you can see the longer lens is the 40x objective (40 times). The shorter one, over the slide is the 4x (4 times) objective. The rule is simple: The higher the magnification – the closer the lens will be to the specimen. In most cases the highest magnification lenses have the longest barrels. Look at the picture below.

That short lens hanging down over the petri dish has the lowest magnification. That long lens is pointed away from the dish. It has the highest magnification. That is the correct way to use the compound scope with a petri dish. Let's say you would like to see a tardigrade at high magnification. That means the objective lens (the longest one in the picture) would have to almost touch the specimen. Only if there is a microscope slide on the stage would that be possible. You could rack the lens right down to

the surface of the slide, almost touching it. You might see the tardigrade's leg only, magnified 900 times. But this would be impossible to do when looking in a petri dish full of water, because the objective lens would be under water. Do not use the higher power objective lenses when looking in a petri dish. This pertains only to compound microscopes. Dissecting scopes do not have this problem.

More about Objective Lenses

Most compound microscopes have a short lens marked "4x" that will give you 40 times magnification with your 10x eyepiece. That is perfectly fine for finding a tardigrade. Therefore, your compound microscope can find a tardigrade with an objective that will not to touch the water. Take a look at some objective lenses in the below picture. You will see their magnification and you can compare how long or short they are. This is important when you want a long working distance between the specimen and the lens itself. The objectives below are 2.5x, 4x, 10x, 20x, and 40x. The little black objective (second from the left) is a 4x, the one just mentioned. Since it is short, it will not touch the water. To the left of this black objective is a much longer 2.5x (marked "F 2,5/"). This one is okay too (even though it is long), because it is specially designed to work far away from the stage (and the petri dish of water). The lower the power (4x or 2.5x), the more working distance you will have.

Your lowest magnification when using a compound microscope will be perfect to find a tardigrade. Tardigrades are about a millimeter in size which is huge under a microscope. Sellers of microscopes tend to stress more power. The advertisements or even the box that the microscope comes in usually will emphasize higher power. This is not always needed. In fact, much higher powers make it more difficult to search through the jungle of lichen or moss where you might find a tardigrade. If you can buy an objective lens of 2.5x (like the one on the previous page) for your compound microscope, this is desirable when searching for tardigrades. Used with a 10x eyepiece, it will give you a low magnification of 25 times. Low power makes it a bit easier to spot a tardi-

grade. The reason is because you will see more in the field of view when you search at lower power. It is like the difference between searching on foot or by airplane for a lost hiker in the mountains. On foot, you must search every bit of the terrain very slowly (high power objective = close up view). By airplane, you can scan miles of terrain looking for your lost hiker (low power objective = wider view). The man is small when you see him from the air, but you can sure find him quicker. On foot, the man is much bigger, but it is very hard to find him. This is why a dissecting microscope is so good for the initial search for tardigrades. It is a low power microscope with a very wide field of view and a long working distance between the lens and the specimen on the base (see above photo of three students using dissecting scopes).

The Eyepiece

To find out the magnification of something you are looking at, you simply multiply the eyepiece number times the objective number. Eyepieces are almost always 10x (10 times). This picture shows the eyepiece of a simple microscope. If you have a 4x (4 times) objective in

place, and you use a 10x eyepiece, your specimen is seen at 40 times total magnification. You might also say at "40 power." Here is the arithmetic again: You

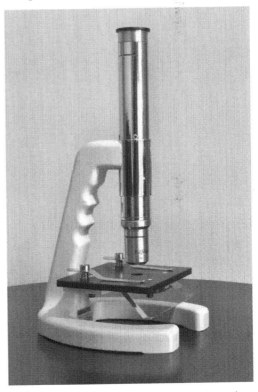

multiply the eyepiece number times the objective number. It's that simple. Eyepiece (10x above) times an objective (say 4x) to get total magnification of 40 power. Likewise a 10x eyepiece times a 10x objective gives you 100 times magnification. These magnifications are more than enough to see and photograph tardigrades.

The Field Microscope

You may want to get a microscope that is portable, something you can take on trips to explore what you find right on the spot. Consider something that uses a mirror or a prism, rather than a microscope that you must plug in to an electrical outlet. The above pictured microscope has a simple design, and uses a prism to gather the light. It is a compound micro-scope, but it does not have a rotating turret with different objective lenses. In this scope, you must screw in a different objective lens each time you want to change magnification. That is the trade-off for light weight portability, and no moving parts. In my recent video on YouTube, "First Animal to Survive in Space," I'm using this prism microscope. I like it because it is portable, and has no built in light. You can use it outside. Since it needs no extra light, there are no batteries or external power required.

Elegant in its simplicity, it is similar to the microscope made of brass that my father had when he was a teenager. It's a tried and true design. It is a great "field microscope" because the prism underneath takes in light from anywhere and bends it up into the objective lens. It has no gears- just a sliding tube to adjust the focus.

Lighting

You will need extra light, and you can simply use an LED flashlight, which is very bright. You can mount the light in a number of creative ways, but the easiest is to use an electronics tool holder. Since you must shine a light across the specimen dish to find tardigrades, that is a good way to do it. Find a small LED flashlight that has an on-off switch or has a screw in bulb that stays on. Then make or find a tool holder to keep it in position. Even if your dissecting scope or your compound microscope has a built in light, you will need your own adjustable light source. You can also use a high intensity desk lamp. Anything that will give you a focused beam of light will work. In the picture on the next page, I'm using a keychain LED light on an electronics tool holder. On the microscope, I'm using a colored slide and a piece of tissue paper for clarity - so you can see how the light is reflected up to where your specimen would be. (Normally, this would be a clear glass slide.)

When light comes up through the microscope it is called "transmitted light." You see your specimen in what is called "brightfield," because everything

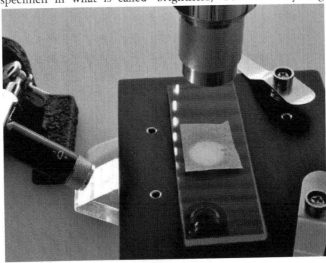

around your specimen is very bright. When you first look for a tardigrade in a specimen dish, you must stream the light low and across the bottom of the petri dish. More about that in the next chapter.

Higher Magnification Problems

After you have first located a tardigrade in your suspension dish, you must transfer it to a microscope slide, and then you may view it at higher magnifica-

tion. To the right is a picture of mold spores magnified at 400 times. Notice how few of them are in clear focus. See how dark many of them are. These are the problems with high magnification. First, you can't keep everything in focus. After the tardigrade is mounted on a glass microscope slide, can you view it at higher powers, but not all of the tardigrade will be in focus. Second, there is way less light coming into your microscope. Magnification is all about looking at a small area and blowing it up in size. Said in a different way, the higher the magnification, the smaller the area is that you see. Because the area is smaller, the amount of light is smaller too. It is not as easy as you think to properly illuminate a specimen when you have magnified it 1000 times. The specimen will be very dark, and you must have strong focused light.

Depth of Field

"Depth of field" refers to how much of your image stays in focus. As mentioned on the previous page, you cannot keep everything in focus at higher powers of magnification. Microscope lenses, unlike your eye, are designed to focus at one level at a time. One part of your specimen is always out of focus when you are focused on another part. At higher powers, you can only focus on one layer at a time. Higher magnification gives a shallow depth of field.

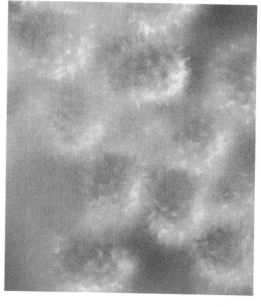

While many microscopes come with a 100x objective (giving you 1000 times magnification), you will rarely use it. Hardly ever do you need to see something at 1000 times magnification. Bacteria and blood cells, yes. Everything else, no. Look at the above sunflower pollen photo taken at 400 times magnification. You can already see that there is less in focus than you would like to see. At 1000 times magnification, nothing would be in focus except a few spikes of one pollen grain, and that is not too interesting.

Real Life Example

A real life example would be the below picture of an elk. It would be like looking at an elk in front of a forest where everything is a total blur, except a one inch slice- his nose. Using a microscope at higher powers only allows you to look at any single slice of your specimen, and only at one slice at a time. In our elk analogy, you have a choice to see clearly either the edges of a few trees, or the tips of his horns, or his eyes, or his ears, or in this case his nose.

But you could not see the whole scene as it looks. To see the whole scene, you need a lower powered objective lens like a 4x or 10x (see picture on next page). That means you are magnifying what you see by only 40 times or 100 times. That's plenty for most tardigrade observations. Tardigrades are rather big under the microscope and you do not need a lot of power to find them, observe them, or photograph them. You could even use a hand held magnifying lens to find a tardigrade in a petri dish of water.

Immersion Oil

Here is one final important piece of information about very high magnification objectives. These lenses (like the 100x) need oil, a special microscope oil, on the surface of the slide and on the front lens of the objective. This oil is called "immersion oil." Here is why. Normally there is air in the tiny space between the cover glass and the objective lens. Light gets bounced around in the air, and some clarity is lost. At high magnifications, up around 1000 times, this can make a big difference in how clear the specimen looks. So the solution is to use oil instead of air. The objective lens of a 100x objective is almost touching the slide anyway. A drop of oil is placed on the cover glass, and the lens is slowly allowed to come in contact with the oil.

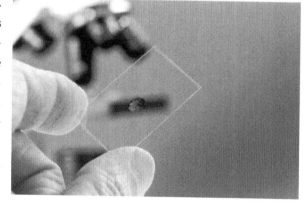

Immersion oil handles light the same way that

glass does. When there is a thin layer of oil between the objective lens and the cover glass, it is the same as if the light was passing through glass the whole way. It is not scattered, and the image is improved. These objective lenses are designed to handle the oil and will not be damaged by it. They are called "oil objectives." The only problem is that oil is a bit messy, and you have to clean it off the slide and off the objective lens. Therefore, you have to question your need for this type of objective. Don't be swayed by claims of high power of 1000 times magnification when selecting a microscope. Stick with basics like 4x, 10x, and 40x objectives which are all you need for most observations.

Now you are ready to actually search the petri dish for some live tardigrades.

CHAPTER 6

Spotting Tardigrades in the Specimen Dish

The night before the science fair, make sure you prepare some petri dish suspensions so the tardigrades have time to rehydrate. Then you can bring those in and set them on your display table, and there is a very good chance that you will be able to show the judges and visitors some live moving tardigrades. Here is the step by step process.

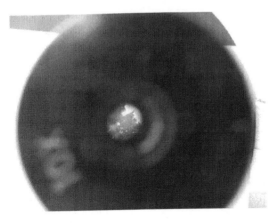

Special Cardboard

To help you see any tardigrades that may be in your specimen dish, you'll need to prepare a special piece of cardboard or paper. You must find some black electrical tape, or a sheet of black plastic. A crafts store sells sheets of black paper, and that will work. Or, you can spray paint a piece of thin plastic or cardboard. I used an index card for my backing. Cut and paste strips of black electrical tape until you get a nice sized black rectangle onto your backing. This should be large enough to cover the entire bottom of your specimen dish. Pictured here is the index card and wide black plastic tape method. You can use a piece of flat cardboard from the back of a note pad too. Black electrical tape is nice, or you can also use the black tape which is sold in craft stores. This also comes in flat adhesive sheets.

Trim the black card backing to the size of your microscope stage and, if necessary, punch holes for the microscope clips. You can use an ordinary hole puncher for this, and then use a scissors to cut slots from the edge of the card in to the hole. This way you can slide the card under the microscope clips (see below picture of this). In the below picture, you see a petri dish with light from the side, all set up with this black card. See how the card fits right under the clips. In this case a compound microscope is being used. A dissecting microscope would actually be better for this task, if you have one.

The reason we use the black card under the dish is because it makes it very easy to spot tardigrades. Against the black background, the tardigrades and other living creatures light up brightly like little light bulbs. This technique is called "incident illumination," and what you see when you look through the microscope is called a "darkfield" view, because everything in the field around your subject is dark.

Lighting

Somehow, you will have to mount and set up a narrow beam of light, from a flashlight or even a high intensity desk lamp. This is where you get creative.

As mentioned in the previous chapter, I use a tiny keychain LED flashlight for this, or a slightly bigger pocket sized LED flashlight. You must have a narrow beam that you can direct at a low angle.

Gently place your specimen dish on the microscope stage and shoot the beam across the bottom of specimen dish (pictured on right). You want to illuminate the suspension from the side. You will see any microscopic creatures stand out like a field of bright stars against a midnight black night sky.

Next, focus the microscope at lowest power on the debris that rests on the bottom of the dish. If you use a dissecting microscope, this will be very easy because you can see the dish properly illuminated from the side with a large distance between the objective and the dish (see picture on previous page).

Using a Compound Microscope

If you use a compound microscope, you must be very careful. Compound microscopes, like the one in the below picture are not designed to be used to search a petri dish, so never use an objective higher than 4x. REMEMBER: If you try to use a higher power objective lens (10x and above), you will wind up with it going under water. You do not want that to happen or you will ruin the optics. You must use the lowest power objective because it gives you a good "working distance" from the surface of the water. Check to see if your objective is very close to the water as you start to focus. If so, you will need to drain some water out of the dish using an eyedropper.

Whether you use a dissecting scope or a compound scope, focus on the bottom of the dish, with the light streaming sideways across, so you can methodically examine the specimen. Make sure you have removed (or moved out of the way) the metal clips for holding microscope slides in place.

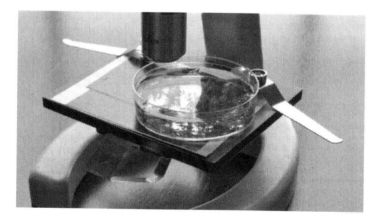

Start at the outside edge of the dish as shown in the above picture (taken with LED light out of the way for photo clarity). If the dish is round, slowly rotate the dish in a circle checking along the outer edges. Then move the dish over either left or right, and do this again, inspecting another circle, closer in towards the center. By using this method, you are covering the entire dish methodically in ever decreasing sized circles.

If you are using a rectangular dish, then start at one end, and move the dish slowly away from you, move it over a step, and then towards you as if you were mowing the lawn, covering strip by strip of search area across the dish.

You are hunting for bears. These are fairly slow moving creatures, which is why they have been named tardi-grades, or slow-walkers. Don't worry; they will not run across the dish while you look at one side, and then run back while you look at the other side. Be confident. If there is a tardigrade in your suspension, you will find it.

What Do They Look Like?

By now, you may already know what a tardigrade looks like from the pictures you've seen. An internet search will show you many more examples. Below is a tardigrade with stripes, viewed in transmitted light. When you first search your petri dish, you will not see this kind of detail. A tardigrade will **not** look this big or this clear during your initial search of the petri dish.

What you will see is more like the picture on the next page. The tardigrade will be tiny, and lit up like a little white light bulb. Looking in your petri dish for the first time will be a challenge, but a fun one. Don't be discouraged if you don't find any tardigrades at first. Sometimes, they can take two or three days to emerge from their hibernation. Sometimes you have to

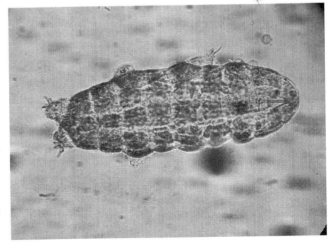

prepare another suspension from the same sample of lichen or moss. Sometimes you have to make a suspension from a brand new sample of lichen or moss. Eventually, however, you will find one.

Part of the process is getting used to looking through the microscope in the first place, positioning the light, learning to move the dish, getting into a methodical search pattern. Search and Rescue pilots and their scanner crews are taught very specific patterns to use when flying the airplane and when searching the ground. This is very similar, and getting used to a technique of scanning is part of the learning curve. You can use two pencils, each with a long sewing needle taped to the end, or two sharp knitting needles to pull apart some of the lichen or moss that clutters up your view.

Your first tardigrade will probably not be strolling across the bottom of the 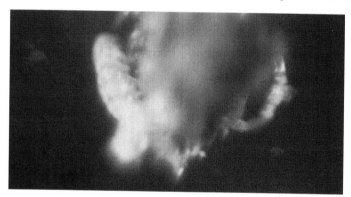 dish in perfect focus. Instead, it will most likely be clinging and crawling on some piece of litter and will be quite out of focus. You will have to recognize it by its color and movement. It might look like the above example. Can you find the tardigrade?

One method I like to use is to frequently hold the dish still, or let go of it momentarily; don't touch anything and watch an area for a few seconds. If you see a piece of debris moving by itself, it means that there could be a tardigrade hiding behind it. Wait a bit longer, and out he pops.

Or your thoughts may be, "Oh. That was just a rotifer. On to the next tiny chunk. Oh, there's a roundworm. You can't fool me with that squiggly movement. What's that? YES. I found one."

You have to be very observant, because the tardigrade may be holding still, just clinging to a piece of debris. It will look like the below picture, moving around, not at all in focus.

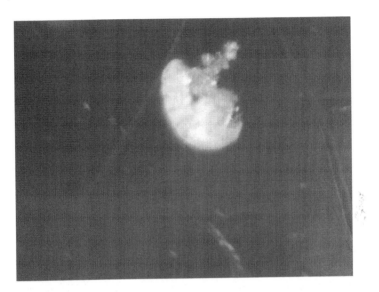

Be patient and persistent, and you are sure to find a tardigrade in the petri dish. Once you do, carefully make a mental note of where it is. There is a saying that your first tardigrade is the hardest to find. It will be clinging onto a piece of debris or vegetation with back claws, making a swimming motion with its front claws.

In the next chapter we will learn how to transfer this tardigrade onto a microscope slide.

CHAPTER 7

Observing Tardigrades at Higher Magnification

If you want to show a close up view of a tardigrade to a visitor to your project, you will need to contain your critter on a microscope slide. You can't chase a tardigrade around the petri dish at high magnification, as explained earlier. Instead, you will find it much easier to drop a tardigrade onto a microscope slide, and keep it relatively centrally located. You will still do a bit of chasing, but there won't be a lot of places for the tardigrade to hide. You can then ask the science fair visitor or judge to take a look, confident that your tardigrade will be either squirming or posing for the viewer.

Slide Preparation

First- clean and prepare a microscope slide and cover glass and have them ready. To avoid lint, use ordinary eyeglass cleaning spray and wipes, or the pre-packaged eyeglass cleaning wipes. I recommend Kimwipes® which are lint free, along with eyeglass cleaning spray. Set this clean slide aside close to you, and have a clean cover glass hanging off the edge of the slide so you can grab it easily.

Using an Eye Dropper

The easiest way to move a tardigrade from a suspension dish onto a slide is by using an eye dropper. It takes practice, and you will surely lose a few tardigrades in the process.

Memorize where the tardigrade is in the petri dish, as you take your eyes away from the top of the microscope. This next part is all done visually with the naked eye, without looking through the microscope.

Use a steady hand by leaning it on the microscope stage slightly and gently lower the eyedropper into the dish, at the edge. Don't go plunging the eyedropper right next to the tardigrade, or it will be gone. After you lower the eyedropper into the edge of the dish – squirt a little air out of it, but not all of the air. See the picture here.

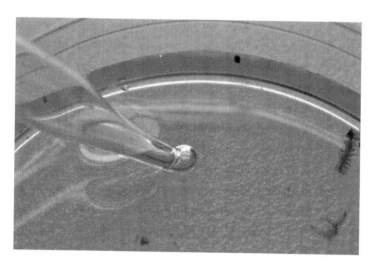

Then suck a little water up the snout of the eyedropper. You must maintain pressure on the bulb of the eyedropper as you suck a couple of drops of water into it. If you loosen too much pressure on the bulb, it will suck in a lot of water. Don't do that, not just yet.

Now, maintaining that steady grip and pressure, slide the eyedropper over to the area near the tardigrade, close, but not right next to it. You should know where the tardigrade is, not even looking in the microscope, because you will recognize the shape of the debris and its location.

In all likelihood, the tardigrade will not be off roaming around, but will still be attached to a piece of plant or debris where you last saw it. That's actually a good thing; otherwise you would not be able to see what you are doing.

In the picture below, the piece of debris we want is directly in front of the eyedropper opening, a bit to the right (dark speck).

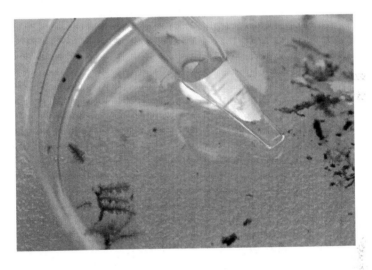

Now, release a bit of pressure and watch the debris, along with the tardigrade get sucked up into the eyedropper. If you loosen too much, you will get too much water in, and you could lose the tardigrade in the eyedropper. So, loosen your grip just enough to snag the tardigrade, and lift the eyedropper out of the water.

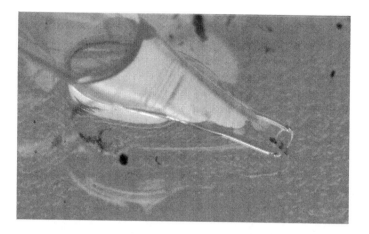

Next, place the eyedropper directly over the center of your microscope slide. You should be able to see the specimen right inside the eyedropper.

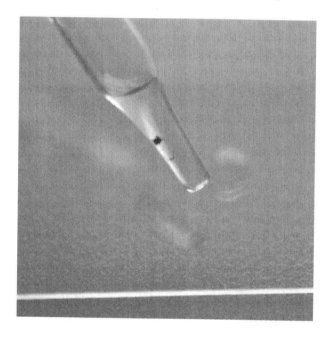

Now, hold the eyedropper vertically for a second so you can see the debris fall to the tip of the eyedropper.

And finally, let out that drop onto your slide. Not all the water; just a small drop. That's one small drop for a tardigrade, one giant leap for mankind.

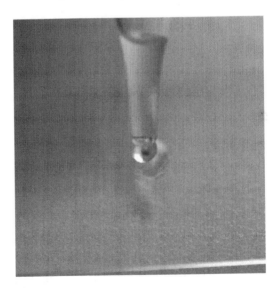

Mission Control: We have splashdown. The tardigrade has landed. It should be on the slide now, right? Better check to be sure. Before you place the cover glass over the drop of water, be sure to check the drop under the microscope, to make sure the tardigrade is in there. It's hard to be certain with the naked eye.

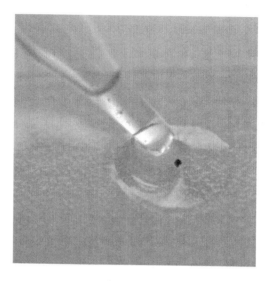

You've done a complex procedure without actually seeing the tardigrade. You have to use your imagination much of the way. Now imagine your surprise though, when you look at the slide and see the tardigrade safe and sound.

It will look something like this next picture. Hey bubble, get out of the way!

It is even possible to do this whole process with a tardigrade alone, without the debris, once you master the eyedropper technique. It's then more like a pantomime performance. Science becomes ballet.

Using a Tweezers

In some cases, you will observe a tardigrade clinging to a rather large strand of vegetation, like a moss frond. In that case, you cannot use an eyedropper.

Simply place a drop of clean water on your slide. Using a very fine tweezers, you can pull the moss strand out of the dish, and place it onto the slide in the middle of the clear drop.

You may find this method much easier than the eyedropper technique. The only disadvantage is that the tardigrade will have a large piece of vegetation to cling to and hide in. Ideally, you will want to isolate the tardigrade and remove all vegetation from the slide, so you can observe and photograph the tardigrade without any debris in the way.

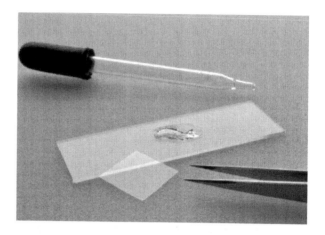

Using your tweezers, pick up the cover glass from the edge of your slide. Then, at an angle, slowly lower the cover glass. The idea is to keep the air bubbles out. It takes a delicate touch, and learning this takes practice. You also need to learn how much water to drop onto the slide. That takes practice too.

It takes a steady hand to lower the cover glass just right and keep the specimen in the middle.

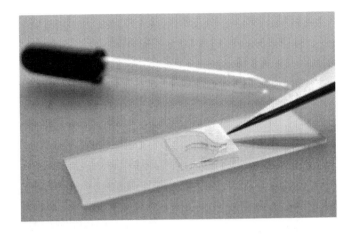

Occasionally, the tardigrade will be shoved to the edge of the cover glass, and you will think it's gone. Check the edges of the cover glass, and you will find it there.

Live on Stage – Your Tardigrade

Now you can place the slide under the microscope, secure the slide edges with the metal stage clips, and turn on the LED light. Initially, I recommend you continue by shining your light from above, and keeping the black cardboard underneath the slide.

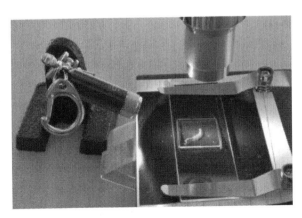

Start by using your 4x objective, and your tardigrade will be big enough to see very clearly in great detail. The light from the top actually gives the tardigrade a natural look, and makes it easy to find.

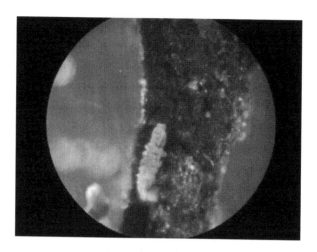

Once you have located the tardigrade with light from the top, you may gently remove the black cardboard. The above picture shows what it might look like.

If you have a 10x, 20x or a 40x objective you most definitely can use these now, but it will be much harder to keep track of your critter because the area you are viewing is so much smaller.

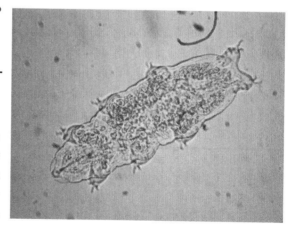

Viewing in Other Types of Light

Now you may also switch the light to below the stage by shining it into the mirror or prism. If your microscope has a built in illuminator, then turn on this light. Ummm...don't forget to remove the black card.

Take your time and try not to move the slide at all. Very gently, just switch in the next higher powered objective. You should be able to identify mouth

parts, see the eyes, claws, and even the digestive system. You can see in the picture on the previous page, taken with green filter and 20x objective, that a tardigrade is almost clear. It would have been very difficult to locate if you started off 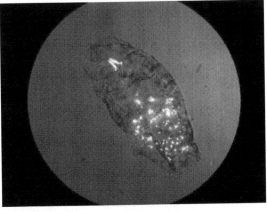 this whole process in brightfield. Since the tardigrade is safely on the slide you can experiment with different types of lighting. Above is a photo of a tardigrade using polarization filters. If you have a more complex microscope, you can use polarizing filters (as shown in the photo above), and you will see the tardigrade mouth parts and digestive system light up brightly. You can also take some creative video while you are set up. You will have plenty of time to enjoy observing the tardigrade on the slide. But wait...

Preserving the Water Level

Water evaporates, and you will notice in the photo on the left that the water line is shrinking under the cover glass. This is not good for the tardigrade, because it means that the cover glass is getting closer and closer to the surface of the slide. This will eventually squash the tardigrade. To replenish the water supply, simply add a tiny drop of water to the edge of the cover glass, and the cover glass will rise to the proper height again.

How can you to leave a tardigrade on a slide for several hours, or overnight? If the science fair is tomorrow, and you are looking a live tardigrade tonight, can you keep it alive? There is a simple way to do this. In the picture below, I used colored components just for the sake of this photo so you can distinguish the parts better. You see a reservoir on the left, a string, and a dark cover glass and slide. Create a small water reservoir that you can place on the slide next to the coverslip (a small bottle cap works). Run a string from the water filled bucket to the edge of the cover glass, and slightly under it. The string will act as a wick, and keep the water under the cover glass at the correct level. The pic-

ture on the next page shows the actual set up to keep a cover glass replenished with water. In this case, I used a bead container for a reservoir, and thick cotton sewing thread for my wick. You can use any type  of thin string which absorbs water. There is no guarantee that your tardigrade will not have escaped, become lost, or climbed up the string. This is a good method, though, to keep water fresh under the cover glass.

Now you know the basics of preparing this water preservation set up. An absorbent string right up against the cover glass leading into a little reservoir of water acts as a wick. This is what you must achieve. If you don't do this, a tardigrade in

this situation might perceive his universe as contracting and expanding. If tardigrades were physicists, they might come up with a theory to explain it, and in this case they would probably call it "string theory."

CHAPTER 8

Mounting Tardigrades on Permanent Slides

Making permanent slides is a lengthy process, though not complex. You will basically have to replace the water with alcohol, and then replace the alcohol with your permanent cement. It can be a bit tricky for beginners.

Pros and Cons of Permanent Slides

Once you have seen a tardigrade in a drop of water on a slide, you may want to make a permanent slide. The cement mounted tardigrade will not look as good as it did when it was alive in water, however, all the essentials of the body will be there as in this picture of a cement mounted tardigrade. You'll probably only be able to see claws, mouthparts, and the outer cuticle or shell.

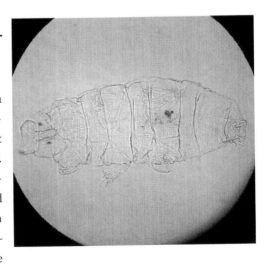

That's why I urge you to take pictures of live tardigrades rather than making permanent slides.

For a science fair or classroom project, you do not need to make permanent slides. Pictures at a science fair are fine. For scientific research or to get a paper published, however, you will need to make permanent slides. Some people enjoy making permanent slides because it is almost an art form and becomes a part of the fun activity in a microscope hobby.

Clearing Agents and Chemicals

A clearing agent is used to get all the water out of your specimen so you can use a non-water based cement to mount your tardigrade to the slide. Some books may tell you to use Xylene as a clearing agent to prepare the tardigrade.

Don't do it. I've tried it, and it is such a powerful solvent, that it will melt practically everything that is not glass or wood. I accidentally spilled some and it immediately melted the plastic petri dish, began to melt my Formica counter top, and everything plastic on it. Xylene is also highly flammable. Do not use it. Instead, I'll describe a safe way I learned to prepare and mount tardigrades. You must remove the water and replace it with alcohol. There are two ways to do this. First, you can transfer the tardigrade in a water drop into a tiny container - smaller than a petri dish, maybe the size of a bottle cap. You could actually use a clean plastic cap from a bottle of soda or a cosmetics bottle. After the tardigrade is resting at the bottom, suck up most of the water with an eyedropper. This is more difficult

than you think, because you could easily lose the tardigrade this way. Then add a few drops of isopropyl alcohol to the bottle cap. This will euthanize the tardigrade quickly, and it typically might curl up into a ball, or go into the tun stage. Now you can mix it up a little, and wait for an hour or so. Using several drops of alcohol in a bottle cap ensures that the tardigrade will have all of the water replaced with alcohol. If you are over 21 years of age, you can use Everclear®, which is almost pure alcohol. Now transfer the alcohol soaked tardigrade to a slide (without a cover glass). Use a magnifying glass or dissecting scope to keep track of your tardigrade and make sure the tardigrade unfolds back to normal size on your slide. It should be fully soaked in alcohol, and the alcohol will be quickly evaporating. *Tardigrade cleared.*

The second clearing method also uses rubbing alcohol (isopropyl alcohol), but you perform the steps right on the slide. This must be done in slow stages. Isolate your tardigrade in a very tiny drop of water on a slide, without a cover glass. To this drop of water, add the tiniest drop of alcohol.

This will euthanize the tardigrade quickly, and as said before, it might curl up into a ball, or go into the tun stage. Let it sit for perhaps an hour, and it will slowly uncurl. You may have to keep adding alcohol because it evaporates so quickly. Wait until the tardigrade uncurls.

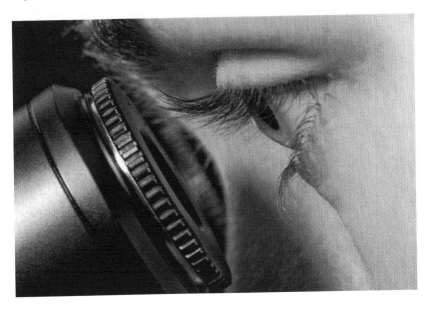

Under the microscope, you can observe it spreading out again into a full sized, deceased tardigrade. Now try to remove some of the liquid because it contains water. What you can do is touch the edge of a piece of paper, or a Kimwipe®, to the water, and let the paper absorb a bit of the water. After you draw off a little liquid, replace it with alcohol. Do this several times until the drop on the slide is pure alcohol. *Tardigrade cleared.*

Those are two ways to clear a tardigrade of water, and have it ready on a slide.

Evaporate the Alcohol

In both of the above methods your goal is to have a fully expanded tardigrade on a slide, without a cover glass, soaked in alcohol instead of water. Next, you must allow the alcohol to evaporate. As you allow the alcohol to evaporate, which will happen fairly quickly, you will need to be ready with your drop of permanent cement.

Mounting Cements

There are very good mounting cements like Oregon Balsam, which is a natural sap from a fir tree (picture above), Hoyer's solution (especially for mounting critters like tardigrades), Cytoseal 60™ for medical specimen mounting, Polyvinyl Alcohol (which is water soluble), and others. Some of these are difficult to obtain and do not work as well as more easily obtained cements. You can use model building glue, and clear nail polish. Even ordinary egg white will work as a good cement for microscope slides.

Some of these cements have their own special process for clearing your specimen of water. For example - say you are using Oregon Balsam. You must use turpentine after the alcohol, because Oregon Balsam is made with turpentine. In this case you would follow the same steps you did to replace the water with alcohol. Only this time, you replace the alcohol with turpentine. It is a lot of work.

If you are going to purchase cement specifically for making microscope slides, my recommendation is Cytoseal 60™, which you can order from a laboratory supply company. It is expensive, though. If you do not want to go through that trouble, you can use plain clear nail polish, or model building cement, although professionals would not use these. Most of these do not need an additional clearing stage. If you are using clear nail polish, you might want to use nail polish remover (acetone) to clear the alcohol. In this case you would follow the same steps you did to replace the water with alcohol. Only this time, you would replace the alcohol with acetone. Again, it is a lot of work.

Fortunately, the simple alcohol clearing is sufficient for most types of glue and cement, if you allow almost every bit of alcohol to evaporate before you drop on the cement in the next step.

Applying the Cement

Just when almost all of the alcohol, acetone, or turpentine has evaporated, at the last moment, you will drop a glob of cement on the slide. Then, slowly lower the cover glass over it.

Whatever solvent remains in the tardigrade will be absorbed by the cement, hopefully without clouding. If you leave too much alcohol or water in the specimen or on the slide, the cement will show a foggy look because too much alcohol or any remaining water will not mix well with the cement.

The best way to avoid clouding or a foggy look is to replace the alcohol with turpentine or acetone depending upon the cement you are using. Clear nail polish works best with acetone clearing. Model building cement, and Cytoseal 60™ will absorb turpentine clearing residue with no clouding.

The cement process itself process is simple. Let a drop of cement fall on top of the cleared specimen on the slide, and then you place a cover slip over it (using the tweezers technique you learned earlier). You are done. Just allow it to dry overnight.

The Professional Way

The professional way to cement a permanent slide and lower a cover glass is actually a bit tricky, but you might enjoy the challenge. Here, we will see a method to avoid air bubbles and keep the specimen in the center of the slide. Instead of getting your specimen ready on a slide – you prepare your specimen on a cover glass. Instead of putting your drop of cement on a slide, place it on the cover glass, holding it by the edges. See the picture on the next page.

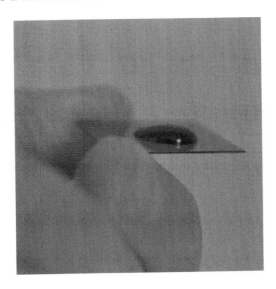

Then, quickly flip the cover glass over so you wind up with the drop hanging from underneath the cover glass. Just like in our eyedropper technique, the specimen is hanging in the exact center of the drip. See the next picture.

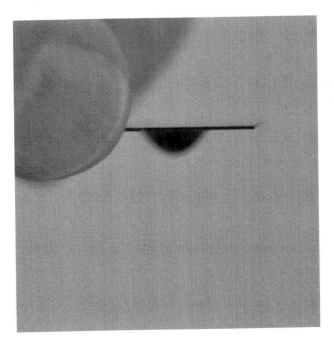

Quickly transfer the cover glass to your tweezers, and gently lower it onto the slide, keeping it level. The tardigrade should be in the exact center of the slide, and all air bubbles evenly sent out the sides.

Double Coverslip Step by Step Process

Making permanent slides is an art form as much as a science, and takes a bit of practice. If you really want to kick it up a notch, here is the *super duper* professional way to make a permanent slide. You use a double coverslip, one round and one square. Here are the steps. I've included some pictures using colored components so you can see more contrast in the photos.

1) Cleared specimen in place on a *round* coverslip;
2) On the round coverslip place drop of cement;
3) Flip over for hanging drop;
4) Lower onto a *square* coverslip.

Let dry overnight (specimen now mounted between two coverslips). Place a drop of cement in the middle of a microscope slide. Lower double cover glass, with *round cover slip facing down*, onto the microscope slide.

Let dry, flat, for a minimum of two days with a tiny 2 gram weight on top to press it down. You will now have a double coverslip, square on top, and your specimen underneath the top layer. The advantage is that the edge of the round coverslip is surrounded by cement, so the tardigrade will be safe in the middle.

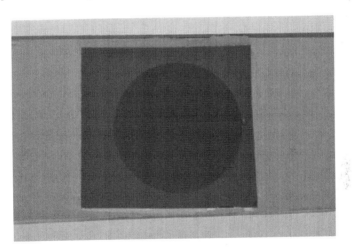

The whole way through the cementing process, you will face the challenge of air bubbles. You have to improve your technique to minimize air, and how you lower coverslips. Even how you handle your cement will improve over time. If you lower a coverslip at an angle, this is one way to squeeze out most of the air.

After you successfully have a tardigrade mounted on a permanent slide, use a fine tipped permanent marker to make a tiny mark next to where the tardigrade is located. This way it will be easy to find under the microscope. To locate your tardigrade, simply find the black dot you made by looking through the microscope. Start at the dot, on low power, then center the tardigrade. Then swing in your higher powered objective lens

You will have a still, permanently mounted tardigrade which you can photograph or make drawings of whenever you wish.

In the next chapter we explore some ways to take pictures of tardigrades.

CHAPTER 9

Taking Pictures of Tardigrades

You must take pictures of your tardigrades for a science fair project, especially if you are not making permanent slides. The display backboard of your tardigrade project should have nice enlargements of photos taken through the microscope. You most likely do not need to do this for a classroom activity.

Above is a science project board about crystals. Notice on the right side that there are photographs of crystals taken through the microscope. There are many types of science projects that can be enhanced with a few photos taken through the microscope.

Taking pictures through a microscope is fun, and you can even use a mobile phone as a camera. In fact, photography through the microscope is a lot easier than you would think.

Type of Camera

The best camera to use is actually a point and shoot camera that has a setting called "macro," with a little flower icon next to the settings button. You simply hold the camera over the eyepiece, and zoom in a bit, and you will see everything you saw with your eye. The camera must allow you to snap pictures with no flash. Most of these cameras let you make videos too, and you can see these on my website and in my YouTube videos.

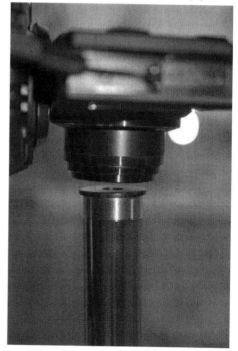

If you plan to take a lot of pictures, you can make a little cradle for your camera. I made one out of PVC plastic pipe fittings, and it worked fine. In Chapter 11, you will learn how to make a camera adapter for your microscope.

Camera Position

For now, let's assume either you've built an adapter, purchased one, or you are using a simple tripod (picture above). If you attach your camera to a tripod and lower it directly over the eyepiece, this will be quite stable.

The main rule is to get the front lens of your camera as close to the eyepiece as possible. In fact, your camera lens should be almost touching the top of the eyepiece. A problem with point and shoot cameras is that after a few minutes of non-use, the camera powers off and the lens retracts. If your camera is on a tripod, simply switch the camera back on. It is already lined up in the correct position. When you turn your camera back on, you are no longer zoomed in, though, and you have a tiny circle of light on the screen. You have to zoom in again to get the full image through the microscope.

No Flash

Set your camera to the NO FLASH setting. That is the lightning bolt icon on the camera. By clicking this, you can toggle to AUTO-FLASH, NO-FLASH, or ALWAYS-FLASH. When you turn off the flash, you allow the camera to determine the correct exposure. Do this for at least one or two shots to test. If your flash were to fire,

the camera would assume that the flash is lighting up the microscope slide (but it is not), and your image will be too dark. If you want to get creative, you can use the FLASH-ON setting and set up a hand mirror that reflects the flash right onto the slide. This actually works great when done correctly. I've used this technique to take pictures of fast moving micro-organisms. The flash freezes their motion and gives the correct exposure.

Macro Mode

Start with the setting on "macro." Use the little flower icon in your camera settings (see picture on previous page). This tells the camera what to do in terms of focusing. Then, experiment and try taking pictures without using the macro setting. Some cameras do not need you to use macro for this type of picture. You can leave just the camera in FULL AUTO mode in many cases. You have to try both settings to know which works best.

Manual Adjustments

You can trick some cameras into a better exposure. For example, you may be able to choose night photography, fireworks, or "dark background" settings. Some point and shoot cameras are more complex and allow you to adjust the exposure manually. If you have any kind of manual override, use it, and try to take some shots with more exposure and with less exposure. There is usually a little plus sign or minus sign to allow manual adjustments. Remember to set your flash to OFF, no matter which manual settings you use. If you have exposure compensation settings of any kind, then bracket your exposures +1 and -1. The picture you are trying to achieve will show a tardigrade is not too brightly lit, where you can still see some detail. Try lighting the tardigrade from the top, using an LED flashlight, and from the bottom using the built in base illuminator on your microscope. You can use a high intensity desk lamp too, to illuminate the specimen.

Self Timer

Use the self-timer if you have one. When you are magnifying a specimen 100 times, you are magnifying your camera shake the same amount. The slightest camera shake from your finger pressing the shutter button will blur a photograph taken through the microscope. If you set the self-timer to 2 seconds, then remove your hand from the camera, the camera shutter will do the work perfectly. You can set the timer to 10 seconds as well if you need more time. The idea is to prevent camera shake, which is magnified as you increase power through the microscope. If your tardigrade is moving a lot, then you can't wait 10 seconds or even 2 seconds before taking a picture. In that case you will want to use a remote release.

Remote Release

Many of the better point and shoot cameras have a remote release, where you can fire the camera using a remote control. This is also a very good way to eliminate camera shake. This remote release plugs into the camera and you can control many of the camera functions by pushing buttons. The one shown here

has a long cord attached. There are also wireless versions for some cameras, and you can be several feet away or even across the room to click the shutter. Some cameras  have a remote capture computer program that you can purchase or that might even be available online for free (or may come with your camera software). You connect your camera to your computer with a USB cable, run the remote capture program, and you have all the controls at your keyboard and mouse. Since you don't need to touch the camera, it also avoids any vibration.

You will be able to instantly snap a picture using your computer the moment the tardigrade pops over the edge of his little piece of plant fiber.

Most remote capture programs have a viewer, so you see everything through the microscope on your computer

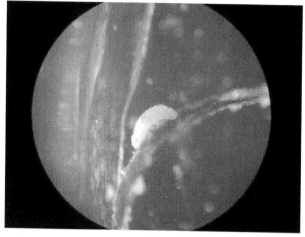

monitor, and you can take a series of quick pictures with a few mouse clicks.

Other Cameras

You now know the main considerations for simple photography through a microscope. Try all of these techniques using a video camera too. Try hand holding your cell phone camera. Since every camera is different, you should

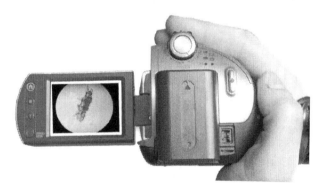

try as many kinds of cameras as you can, to see what works best. Of course, if you have a digital SLR (DSLR) camera, there is no doubt you would try shooting stills and video through that too. You will find DSLR's very challenging though, unless you purchase special adapters for your camera, as series of rings and tubes, which allow photography through a microscope or through a telescope.

Making Prints

Making prints for a science fair display does not have to be expensive. You can print your digital images at home on your own inkjet or laserjet printer.

This uses up a lot of ink, and you might want to consider having your prints made at an office supply store. There, you can simply bring your SD card, and they will make large prints on 8 ½ x 11 inch paper for you. This is the least expensive way to make prints. You could also go to the photo-finishing department in the large drug store or department store,

but those prints get expensive when you order sizes larger than 4 x 6 inches. When you do order 4 x 6 inch photos in the electronics or photo department of a big store, these are usually very affordable, and large enough to see well on your science fair display board. If you are a teacher, you could make these prints as well, and hand them out to students to pass around or take home. If your school has a color printer or a media center, then you could make color prints right from a computer to the color printer. Some printers also have an SD card slot that you can use.

For your science fair display board, you can attach prints of all sizes. The below picture is of a science fair display about which soils were best for plant growth. There were a few microscope photos pasted to the board (upper right). Make many prints, and cut and trim them for your display.

Now that you know all of the basics, we can move on to doing a science project using all we have covered.

CHAPTER 10

Doing a Tardigrade Science Project

Finally, we are at the point where we have the knowledge needed to start a project with tardigrades. Whether you are teacher who would like to set up a classroom project or a student entering a science fair, this chapter will provide some good ideas for both purposes.

You can do a variety of experiments with tardigrades well beyond those suggested here. Use the suggestions here as a beginning to spur your own imagination. If you e-mail me some of your ideas, I might use them in the next edition of this book.

Further, this book cannot address in detail how to conduct every type of science project, so I recommend you refer to some of the many science fair books available. I've listed a few of these in my Amazon store (link at the back of this book).

What is advantageous about using tardigrades in a science project is that they have some unique characteristics, which can make your project stand out in a science fair. Here are just 5 great things about a tardigrade science project:

1) Tardigrades are big compared to other microorganisms, so when you find one, they are easy to see at lower powers and easy to photograph as well. If you have a live tardigrade at your presentation, anyone can look into your microscope and see it easily.

2) Tardigrades can withstand temperatures at both extremes, so you can conduct a variety of experiments testing what their limits would be in various environments: the arctic, the desert, the planet Venus.

3) Tardigrades move slowly, tardi (slow), grada (stepper). That makes them easier to work with than most other microorganisms, which race around the petri dish or slide. You can actually make good drawings of them from your own observations because they are so slow compared to other micro-critters.

4) Tardigrades can be easily identified. You don't need many books or years of study to know what genus or species of tardigrade you are seeing.

5) Tardigrades are everywhere. Have no fear, tardigrades are here. If you want to do a science project about them, you don't have to buy them from an online company and wait for them to arrive in the mail. Go outside and get some!

Five Steps

Doing any science project involves the following five steps, or a variation of these:

1) Settle on your idea. Teachers and students should bounce ideas around. When you finalize your idea, form it into a question or concept you can test. This will allow you to do a project about a single concept, testing your hypothesis. A hypothesis is a statement or a question that you can verify or disprove.

2) Do some research on the internet, search the library, and check Amazon and Kindle to see if there are references you will be able to use to support your project idea. Search magazines, journals, and online documents. See if you can meet people who are familiar with your research such as teachers, university professors, or

experts in the field. Do interviews on the phone or through e-mail. Note all of your sources so you can give credit to them.

3) You will have to design the actual test or experiment. This is an investigation, and it also explores what factors may influence the investigation. These factors are called variables. Designing the experiment is the most difficult part of the project. Doing the project is the fun part. Thinking it out before you start is hard. Students at Scottsboro High School, in Scottsboro, Alabama designed a very complex tardigrade project to determine if life could exist on the planet Venus. Their project design involved 80% spacecraft payload design considerations and 20% tardigrade research considerations. They stated the benefits of their project to be both a study of Venusian atmosphere and a test of the theory of life on Earth being of extraterrestrial origin.

4) Collecting the data. This is done using tools like a thermometer, a microscope, camera, measuring cups, GPS, and so forth. How would you collect data from another planet or from space? You would need microscope cameras and transmitters. Collecting tardigrades involves care to prevent contamination, and requires consistency in the collection techniques. Even eyedroppers can contaminate, so it is better to buy a dozen inexpensive glass eyedroppers, or a hundred plastic disposable ones, than to use the same eyedropper and not know where your tardigrade really came from.

5) Analyzing the data. After the experiment is over, you will state your observations and suggest some possible conclusions. This is done with charts, pictures, graphs, and tables, as well as with written description. At this point it is important to relate these conclusions and the analysis back to the original hypothesis. That is what it is all about. Was the hypothesis proved or disproved? What caused this conclusion? Is there anything we can learn from it that has an impact outside this experiment? If there are few or no tardigrades in the lichen of trees surrounding a popular

campsite with a wood fire pit, does this mean that tardigrades would be affected by pollution from a factory?

Now we can insert the above 5 step process into the context of setting up a science fair project. Let's review the typical elements of a science fair project with a view towards how you might incorporate tardigrades into each aspect of your project.

Presentation

Your project is presented on a table with a large display board behind it. Since it is fairly easy to take photos with even a mobile phone camera, you will surely have some great images on your display board. On the next page is a picture of the display from a sunflower project.

In the sunflower project display, notice how nice the micro-photos look on the lower right side. All taken through the microscope, they make any science project stand out. After you make your prints, cut them out in circles – so they look like images you would see through the microscope. For a template you can use a soup can or margarine container. Anything circular of a large size will do. Trace the circle on your picture, and just use a scissors.

Abstract

This is a summary of what the project is about. Keep it simple, but give yourself credit for the tough stuff by mentioning how many types of trees, how many samples collected, and types of tardigrades found, and the results. This can be a small printed paragraph in the center of your display. An abstract is short, about a paragraph or two. It answers the question: What's this project basically about?

Hypothesis or Question

What is the project trying to prove or find out? Typical projects might have types of questions like these: Which types of popcorn pop best? Do plants grow in the dark? A tardigrade project might ask: Can tardigrades survive the cold or dryness on the planet Mars? Can tardigrades survive in carbon dioxide? In pure oxygen? What do tardigrades eat? The hypothesis is clear cut, and allows for a reasonably certain answer. The hypothesis takes into consideration that there are methods that exist to work towards exploring an answer. It is pointless to set up a project with a question that cannot be reasonably answered. Does life exist on Mars? That is a more reasonable question than: Does life exist in the Andromeda galaxy? We know many things about Mars that might make life possible or impossible. We do not have enough information at present to even hope for a reasonable answer as to the possibility of life in Andromeda. In the former case we can do an interesting science fair project, however in the latter case all we could do is present some statistics.

Research

What steps were followed to conduct the investigation? What was actually done to achieve the result? Notebooks are an essential part of any scientific research, and to collect tardigrades for a science project requires you to use a notebook and record some of the following information: What type of material is collected? Lichen, moss, or other material? You must write down date and time of collection, type of tree if you know it, GPS location (easy to see in your smart phone or in a car navigation system), weather conditions, air temperature, and assign a specimen number.

Whether you wish to do university quality tardigrade research, or you want to do a high school science fair project, you will have to be very methodical and document all of it.

You'll need to write down everything you do in a scientific notebook. This can be a simple marble notebook sold in the school supplies section of your local store. Get the kind that does not have pages you can tear out. Do not use a spiral notebook, or a note pad. Get the type of notebook (these are inexpensive) that has pages bound with stitching.

Make sure there are some photos in your notebook of you doing the actual work. See the picture on the next page.

You can write in this notebook, draw in it, and you can paste photographs in this notebook as well. If you make a mistake, you may cross it out, but you should not remove any pages. Judges at a science fair like to flip through this type of notebook and see your drawings, hand written notes, mistakes and successes, and especially photos popping out showing all the hard work you did.

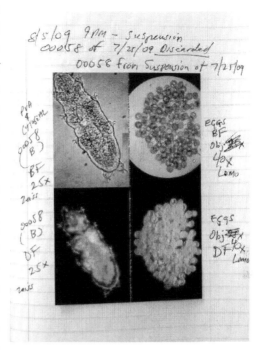

When I did my tardigrade research in New Jersey, I visited every county in the state, collecting specimens. This type of science project is called a population survey. I was neither a student nor a teacher. I just decided to do it on my own, as a private individual, and it took me more than 5 years to complete the project. In the back of this book is my entire scientific paper in case you want to do a similar population survey project in your state. You can download the entire original paper itself from my website.

Materials

What are all the things needed to conduct the research and do the experiments? In order to see a tardigrade you must use a microscope, and bringing in a microscope for visitors and judges to look through makes your project interactive. Nothing shouts science more than a microscope, some samples of moss and lichen in clear jars, and some petri dishes on your display table. Bring to the science fair not only the display board, but all of the materials you used. In the below picture you see part of the display table for a science fair project on The Giant Sequoia Tree.

The table may seem a little cluttered, because it was too small for this project which included a potted live seedling about eight inches tall (far left in the white pot). Some of the other items included on the table were: 6 blocks of other types of wood for comparison with a photo cube showing micro-photo cross sections of these various types of wood (left), a stack of handouts about the project so judges and other students could take away some knowledge, a Sequoia sprout growing in water (right), photos of Sequoia pinecones and an actual Sequoia pinecone and its seeds, samples of lichen that grows on a sequoia, and the student's scientific notebook (far right).

For a tardigrade display, you will want to show models of tardigrade eggs and have a model of a tardigrade. Here is how to create these models.

First, let's look at a few real eggs:

TARDIGRADE EGG

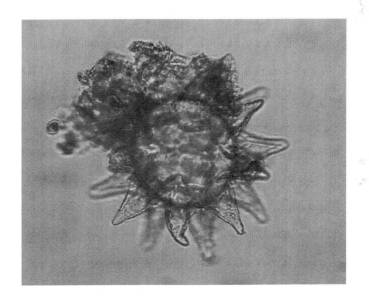

TARDIGRADE EGG

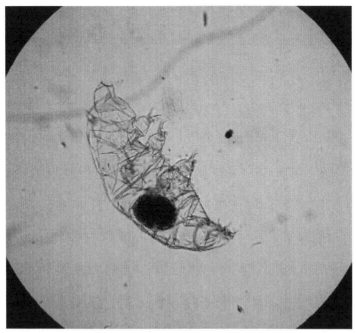

Tardigrade egg in molted tardigrade husk, also called cuticle. Photo taken using brightfield technique.

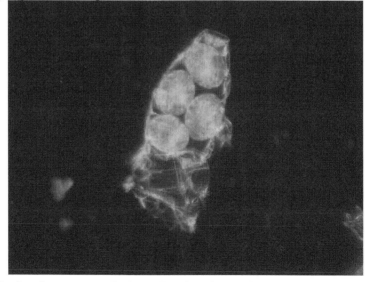

Tardigrade eggs in molted cuticle. This photo taken using darkfield technique.

EGG, BACON & CHEESE MUFFIN.

Okay, that was a joke. Hahaha.

You can obtain realistic models of tardigrade eggs in a crafts or toy store. If you wish, you could spray paint them white. If you leave them as they are, people can pick them up and handle them, although you don't want people throwing them around.

There are other kinds of tardigrade eggs as well: Smooth, with little upside-down champagne glasses, and those with little saucer shapes on them. To represent a smooth tardigrade egg, you could use a fake plastic orange, which has the shape and texture. You could spray paint it white if you wish.

For the two other types of eggs, buy thumb tacks and push pins in an office supply store. From a crafts store, in the floral section, buy a few white two inch Styrofoam balls.

You will need to select out all of the white pushpins, and you will need to spray paint the thumb tacks white.

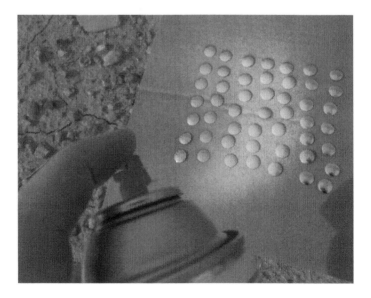

Now start inserting the pushpins and thumb tacks into the balls. They will stay in place for a display, however I recommend you use Elmer's ® School Glue under each tack and push pin first. This way the egg can be handled by spectators and judges. You don't want kids playing with the tardigrade eggs, and having a sharp pushpins falling out. You cannot use any types of glue that have solvents because those will melt the Styrofoam. Use Elmers® only.

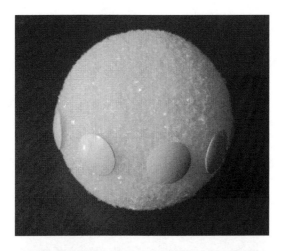

The completed set of tardigrade eggs will look like these below. Do not attempt to spray paint the Styrofoam balls because the paint will melt the Styrofoam.

Now – how to make a tardigrade model. Buy some Fimo® clay in the color you want to use. I like the pinkish color for tardigrades. Some tardigrades are brown, and some are even red. So it is your choice. For this type of tardigrade, we are using pink. Fimo® clay bakes in the oven and becomes hard. When your model is done, you will bake it at 230 Degrees (F) for 30 minutes.

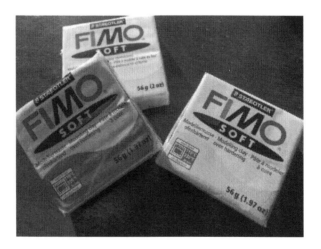

Roll out two logs of clay, one thick and one thinner. They should look like this, and then chop the thinner log into four sections. I use a small cheese spreader as a tool. I work on a piece of hard clear plastic as a surface.

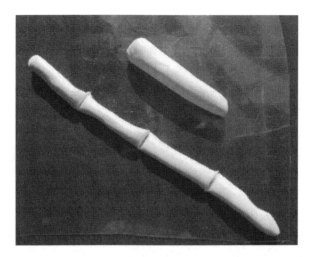

Next, squeeze and roll the ends of the four segments to look like little rolling pins (these will be the legs), and chop the larger log up into 4 body segments.

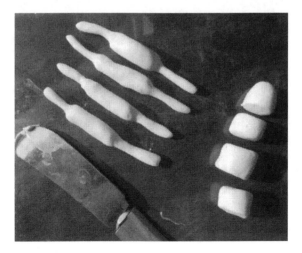

Now, fold in half each of the leg segments, and wedge in-between the body segments. Don't worry about the legs being too long for now. It's called evolution.

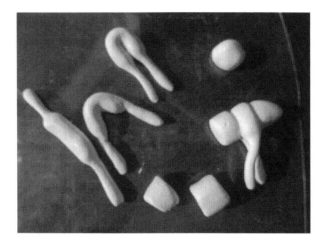

Complete the rough tardigrade, using the back leg segment for the last part of the tardigrade. Just fold it in half and attach it.

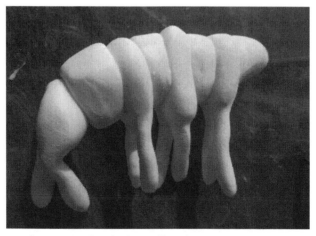

Using a small scissors, clip the legs off, making them short and stubby. Do not smooth the clipped edges. They don't look as good when they are smooth, but that's just my opinion.

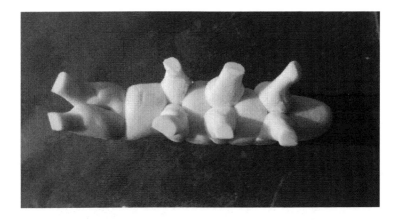

Extend the mouth of the tardigrade, by attaching a piece of clay or just shaping the front end, into a long snout.

Now chop off a piece of the snout, and drive a ball point pen into the snout. Make sure the pen point is retracted.

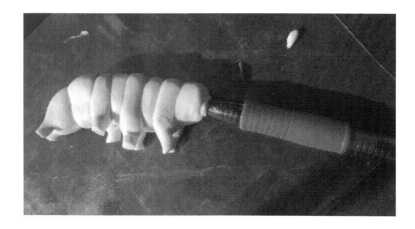

When you remove the pen, you will have a perfect tardigrade mouth.

Now we have to insert the claws. Borrow a scrub brush with long strong stiff white bristles. Cut off at least 24 of them, from different parts of the brush so it won't be noticed. Hahahah.

Insert four of these bristle claws into each foot of the tardigrade. At this point, do not handle the tardigrade too much, or you will distort the shape. Just leave the tardigrade on its back, and insert the bristles. Push them **straight** in, because if you go in at an angle the bristle will poke through the side of the foot.

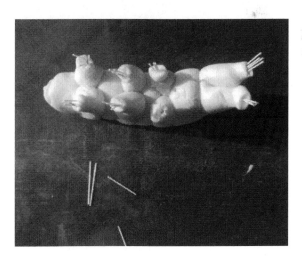

Preheat the oven to 230 degrees F, and when it is ready, put the tardigrade into an ungreased baking pan. Now is the time to make any last minute adjustments to its shape.

Bake for 30 minutes. Using pot holders, remove the pan and let it cool.

Now you can use a jeweler's tool, needle nose pliers, or the round edge of a small screwdriver to curl the claws of the tardigrade so they face backwards. A real tardigrade body has segments, and the segments in your model will look very realistic.

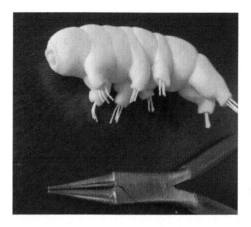

Finally, use a nail clipper to clip the two back claws on each foot, just a bit shorter than the two front claws on each foot. Each foot should have two long and two short claws, hopefully curled a bit.

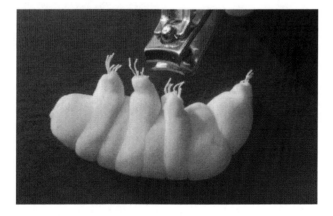

Your tardigrade model is done. Magnificent, right? And you can now brag that you've clipped a tardigrade's toenails!

Procedure

What exact methods were used with all of your materials to do the experiments? All tardigrade projects share common procedures in collection and observation. Since collection and observation is what this book emphasizes first and foremost, this book makes the hardest part of the science project easy for you now. You can literally use the lessons in this book as a guide to the procedure and as a way to document what you did in your science project.

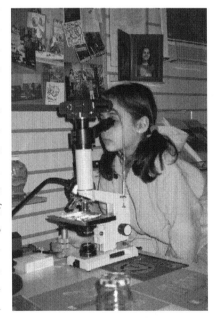

Don't hesitate to have someone take pictures of you carrying out the procedures, looking through the microscope, scraping the lichen. All of these go into your scientific notebook for the judges to see. If you are okay with it, you can also add to the display board some pictures of yourself doing the work. See the center section in the sunflower project display board previously shown (pg. 97). Write down everything you do. That documents the procedure. What brand of bottled water? Get a map showing where you collected your tardigrades, and put yellow dots on it to mark the locations. That would look great on a display. There are many interesting parts of the procedure.

You have plenty of work to do in every stage of the process. For example, you must use a GPS to note the coordinates of each location where you collect your specimens, you must prepare your own permanent slides (if you want to use these), number your collection envelopes (see photo below) and match them to the corresponding slides.

Digital photography is basically free, so take lots of pictures and then decide what you wish to print. Again, document everything in your scientific notebook. Make drawings too, as  these add a personal touch to your observations.

The procedural work is outlined in the first part of this book, but you have to be the one to actually do the work.

The core of your project, the science experiment, is the easiest part. Experimenting with the tardigrades can be very simple. Get some dry ice, place your tardigrades in an old pizza box with the dry ice, and wait a few hours. Did they live? Carbon dioxide may have suffocated them, or they may have frozen. Easy as that experiment was in duplicating the conditions on the surface of Mars, your project will be quite impressive because of all the work involved to find, identify, manipulate, and photograph the tardigrades for your most simple experiment.

Results

What happened? What was observed? Keep this section simple, sticking only to what was observed as a result of the experiment. Don't drift off into conclusions and other ideas. The results section is about describing things.

Maybe in your results section you will describe the types of trees from which you obtained your specimens and how many tardigrades were found in each tree type. That means you will have to collect leaf samples and press them. In Chapter 12 you will learn how to make a plant and leaf press.

Here, you see a pressing of leaves taken at one survey site. Later, you can

look up the leaves in a book on trees and identify the type of tree from which you took your tardigrade sample. The leaf pressing on the left identifies the type of tree, showing the date and the sample number.

It is very easy (once you have a pressing) to make a beautiful color photocopy, and then put that up on your science project display board.

Making pressings have applications beyond just a tardigrade project. Once you build your own leaf press, you can do many types of science projects involving trees, leaves, flowers, seasons, grasses, crops, and more. There are some very inexpensive books on trees and plenty of books in the library on tree identification. You will cite those books in your reference section in the back of your notebook.

In the results section, you also would need to identify the types of tardigrades you find, by studying some of the scientific literature.

Conclusion

This is where we answer our original question: Did we prove or disprove our hypothesis? Can we add anything learned to better the world, or help people, or save the species we are investigating? This is the place where you are allowed to discuss possible impact your project could have outside of the science fair itself.

Now here are some ideas for science projects with tardigrades. As suggested earlier, don't limit yourself to these ideas. If you think of another on your own, then go with it, and share it with me after your project is complete.

10 Tardigrade Project Ideas

1.   What are the body parts of a tardigrade? If some of the below projects seem too complex, you can start in a simple way with this project. Then you can decide if you are ready to go further, or do a more expanded project next year. Each tardigrade project requires you to make some attempt at identifying the genus and/or species of the tardigrades you observe. So that's what this project is about. To make an identification of tardigrade type, you will need to

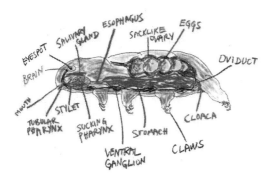

learn all the body parts. Claws, body color, mouth parts, eye location, and internal organs all play a role in tardigrade identification. A simple, but very good science project can be done on this subject.

2.   Can a tardigrade be an environmental indicator of pollution? If you live near a source of pollution such as a chemical plant, power facility, a coal burning power plant, a nuclear site, this will help in setting up an experiment. This type of experiment won one school age researcher the Young Naturalist Award for 2001. You could also set up similar conditions in your back yard using a barbeque grill. By sampling lichen from trees downwind from the pollution source, and as a con-trol group upwind from the pollution, you could make the comparison as to the abundance or non-existence of tardigrades. If you are measuring ground water pollution, you could use tardigrades found in moist leaf litter and mosses in the polluted area, and for your control group, similar sample could be taken in non-polluted fresh water areas. Find leaf litter near a fresh running stream.

3. How slow or fast does a tardigrade walk? This seems like a relatively simple idea, however you have to construct a microscopic measurement system. If you use a ruler, you will have to keep the tardigrade walking in the right direction. Not easy. One approach would be to construct a series of concentric rings on a microscopic level, each ring a known distance apart. Place the tardigrade in the center and measure the time it takes to move from

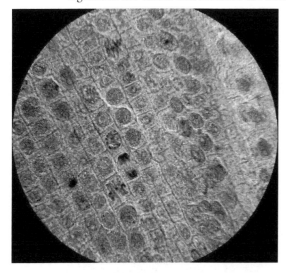

the center in any direction outwards. How to make microscopic rings? Maybe you could use a cross section of a plant, a carrot, or piece of onion skin, as shown on the left, or other natural formation which has microscopic lines or circles. If you can get the tardigrade to move across it, and then measure the time and the distance, you will have the speed. You can even use a clear ruler that has millimeter lines on it. Here is a picture taken through the microscope of a mm ruler. Measuring on the microscopic level is done with a reticle eyepiece and a stage micrometer. But you can use a clear ruler that has millimeter measurements

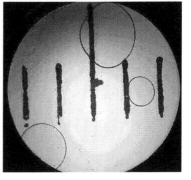

too. Does this seem like a silly project? I don't know of any scientific papers published on this question. If you do your project correctly, you could publish and have your paper cited by scientists around the world whenever that topic comes up in their research. In doing this project, you actually can springboard into another project on measuring tardigrades.

4. How small or large are tardigrades? As briefly described above, you would have to learn how to measure on the microscopic level and this is a very instructional type of project. To add some detail, an eyepiece reticle has a row of tiny lines across the view. These lines do not mean anything by themselves.

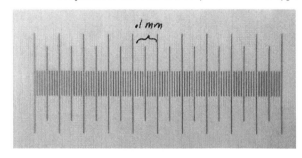

Therefore, you must correlate the distance between two of these lines with a known distance on the microscope stage. To do this properly, you need a stage micrometer (pictured above). This is a slide with ruled lines across it, measuring 1mm. But you could also look at a clear ruler with millimeter lines on it (left). You would easily see how many lines the tardigrade takes up on the ruler. Or by looking at a stage micrometer slide,

through your reticle, you can more accurately see how many lines in the eyepiece equals 1mm. Now you can measure anything you look at under the microscope, such as a tardigrade, by counting how many lines in your eyepiece reticle it measures. By observing a number of tardigrades, and determining their species, you can measure each type and come up with charts and photos for a very nice science project. You have to be careful to use adult tardigrades, so you are not measuring babies thinking incorrectly that they are adults. The way you could ensure you are looking at adults is by allowing isolated tardigrades to remain in the petri dish for several days and explain this in the results. This leads us into another project about tardigrade growth.

5.   How fast do tardigrades grow? Once you have learned how to measure on the microscopic level, you would need tardigrade eggs, and you would have to isolate them and watch them hatch. Then you could take successive measurements over a period of weeks and document your findings. If you are able to observe a tardigrade egg hatch, in an isolated environment, you have another complete project you can do about the relationship between egg type and species.

6.   What egg type is related to a given tardigrade species? In quite a few of the samples I have looked at, I found tardigrade eggs in addition to tardigrades. Although I have photographed both eggs and tardigrades, I never had the time to isolate the eggs and consist-
ently watch them until they hatched and then identify the species. I had one instance where I watched eggs get bigger and bigger as the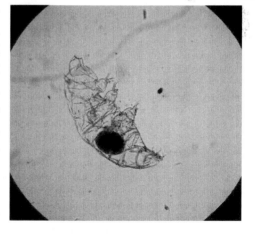
tardigrades inside grew. And after leaving the microscope for only

an hour, I came back and found all the eggs empty. Yes, there were tardigrades crawling around the petri dish. Were these the tardigrades that came from the eggs? I could not be absolutely sure because I had not set up the experiment. I had not isolated the eggs in a petri dish of their own. To document and photograph a well-controlled experiment like this would be a great science fair project, and also contribute to the tardigrade knowledge we have.

7.   Does cryptobiosis extend lifespan? You can easily measure how long it takes for a tardigrade to rehydrate and come out of its state of suspended animation. Let the tardigrade dry out on the slide, take pictures of it in the dried up state called a "tun," and then just add water. You will have to do some reading research on what cryptobiosis is, explain it, and that can be added to your project as long as you give the references to your sources. By performing this process several times with the same tardigrade, over a period of several weeks, you can prove that you have extended the life span of the tardigrade. If you have time, you can collect samples and save them in envelopes over a period of a year. Do this project a year later and you have shown that the tardigrades have been in cryptobiosis for a year. I still have my original samples collected ten years ago. I could rehydrate the specimens in one of my old sample envelopes and do such a project today, showing a documented increase in lifespan.

8.   What do tardigrades eat or what eats tardigrades? Simple as this project might seem, it may be very challenging. Finding and containing a tardigrade is easy, however isolating it and feeding it is another. You actually would have to provide different types of meals, and document observing the tardigrade eat. Chances are the tardigrade will not be hungry when you are watching it. It might be easier to see what other creatures eat tardigrades. Small insects are very easy to find and gigantic to see under the microscope. You might be able to introduce a tardigrade and see if it is eaten. There are large micro-organisms such as rotifers or anne-

lids like the one pictured here, that may eat tardigrades. Or tardi-

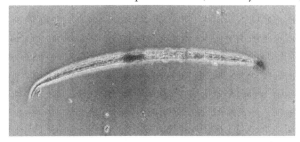

grades may eat them. Find out. The only way to do this is by long hours of observation and experimentation.

9. Are tardigrades carried by the wind? You could take scrapings from a variety of surfaces. You can add another element by sticking to each of these site surfaces something that might be better at capturing tardigrades. I would use self-adhesive Velcro® strips. These come in rolls or packages of small strips, and one side has tiny little hooks, while the other side has fuzzy fabric. Next, put both sides of the Velcro® strip (separated of course) on steel poles, metal signs, wooden fences, brick, stone, aluminum siding, windows and other surfaces exposed to the outdoors, and facing all directions. Include trees that have lichen, putting Velcro® on those trees as well, so you will at least have some comparisons and you will find some tardigrades in the lichen. Leave your Velcro® strips out for a few weeks. Take pictures of the placement process, document with GPS the locations, and take notes of the temperature, time, date and weather conditions. When ready, collect the strips, putting them in individual envelopes, numbered and correlated to the locations. When collecting the strips, this is when you also scrape the raw surfaces close to where you had placed the Velcro®. Scape the metal, wood, brick, glass, and of course the lichen from the trees, and correlate these samples also, to the same locations as the Velcro® strips. Now you can see if tardigrades are found in the two types of Velcro® surfaces, or on the bare surfaces, on neither or both, for every location. We know that tardigrades are found in lichen or moss. Is that because of nutrients, food, in those habitats? Or will tardi-

grades collect in Velcro®? If they do collect on any surface, can we conclude that they are carried by the wind? This is something that has been little researched in the field of tardigrades.

10. How long does it take a tardigrade to revive from the tun stage? This is relatively easy to determine. You simply allow the water on a slide to evaporate while you watch the tardigrade shrink and go into the tun stage. Leave the slide out overnight, covered of course so your tun doesn't roll away or blow away. Then get out your stopwatch, and just add water. You can do a

video or take a series of photos of the rehydration. When the tardigrade is fully mobile again, you have the exact time it takes to rehydrate out of cryptobiosis. I would do this with several different tardigrades, preferably of different species. Alternatively, you could do this with all of the same species, and then draw the conclusion that this particular species averages such and such a time to rehydrate. That is only rehydration. What about freezing a tardigrade on a slide, freezing the water too? Then defrost it, and document the time to defrost. This would make a nice comparison to drying out and rehydration at room temperature vs. freezing and defrosting to room temperature. Adults might want to use Dry Ice or Liquid Nitrogen.

Doing a science fair project about tardigrades can be very challenging and lots of fun. It will definitely teach you the basics of good observation and recording. Who knows? You may even win first place!

CHAPTER 11

How to Make a Microscope Camera Adapter

T he best place to buy your parts is at the hardware store. The big chain stores or even your locally owned store should have everything you need. The adapter does not have to be fancy or expensive. It just has to be more stable than your hand held camera.

Type of Camera

This chapter is intended to help you make an adapter for *a point and shoot digital camera, not a DSLR*. A heavy DSLR (Digital Single Lens Reflex) like the one on the right really should be mounted on a tripod, and then centered over the microscope eyepiece. Most microscopes are not really designed to support heavy equipment sitting on top. Besides, DSLR's have threaded or bay-

onet fronts, on which you can attach commercially made adapters. Since you would have to purchase an adapter specifically for your DSLR camera model or lens, and possibly a series of step rings in between, that's *not* what this chapter is about.

Adapters

Let's look at traditional adapters first. To the right is a picture of a commercially made microscope adapter for a DSLR or for an older 35mm SLR camera. The bottom part has the thumbscrew which tightens on the barrel of the eyepiece tube. The top is threaded for adapter rings which will fit the camera, and there is an extra eyepiece out of the side. It is a very expensive type of adapter.

To the right is another type of adapter, less expensive, however still for DSLR or SLR only. Here the camera uses the lens of the eyepiece instead of its own lens.

The good news is that the simple point and shoot adapter you can make yourself may possibly work for your DSLR, if you really want to forego buying a commercially made adapter and the step rings (like the one on the right).

A point and shoot digital camera typically does not have anything on the front, threaded or bayonet, onto which you can attach a stable microscope adapter. This is because point and shoot cameras are meant to be compact, and the lens retracts into the camera body leaving a flat surface. So the idea is for us to build an inexpensive adapter. You will simply drop your camera into it and have the stability you need.

Above, the early Canon A40 point and shoot camera seen here did indeed have a bayonet socket which is surrounding the lens. You could attach an adapter ring on the front and then you could connect it to a microscope stepdown ring. But nowadays, you probably will not find this kind of manufactured adapter or point and shoot camera. You will have to make your own adapter.

In fact, with an inexpensive homemade adapter, you will be able to shoot great microscope pictures with not only an inexpensive point and shoot camera, but also with a lightweight DSLR, a webcam, and even your smart phone or tablet.

My camera for the above set up is a Sony Cyber-shot 7.2 mega pixel point and shoot. The image is of the slide with the piece of moss which was used in the previous pictures.

Test Your Camera First

Before you begin buying the components for your adapter, you need to make sure your camera can take pictures through your microscope. Start by setting up a nice colorful specimen on a slide, perhaps a strand of moss, or a strew of colored sugar crystals. Try to cover the whole field of view in the microscope from edge to edge with your specimen. Focus the microscope and make sure you have a perfectly sharp image holding your eye up to the eyepiece. On the facing page, I'm using a Nikon Coolpix S8200 for the test just described.

Turn on your camera, put the setting on macro, and either set it up on a tripod or hand-hold it over the eyepiece. See what kind of image you get, and then if need be, zoom in a bit. Try to zoom in until the circle of light fills the entire camera LCD screen with your specimen image. If you can only get a small circle as in the below picture, filled with a decent image of your specimen, then that is all your camera can do. Consider buying another camera if you want a bigger image. Note: A more expensive camera will not necessarily give you the best image in the view finder. It is more a question of optics than a question of price.

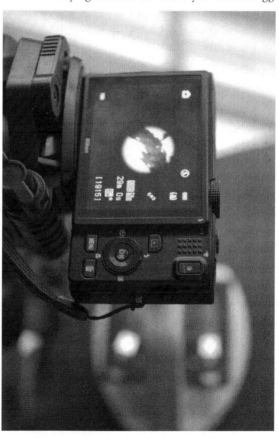

When testing your camera, hold the camera as close to the eyepiece as you can, without actually touching the eyepiece. It is best if you use a tripod to make this work (as I did here with the Nikon S8200). I slowly lowered the camera into place, keeping the image in the center of the screen. Now take a measurement of the ideal distance between the front of your camera lens and the top of the microscope eyepiece. It should be only two or three millimeters. See the picture on the next page. If the distance is too small to measure, then that's okay. Just make a mental note of it.

Measuring the Eyepiece and Camera

Microscopes have standard eyepiece and tube diameters on the outside. That's all you need to be concerned with. The outside diameter is usually about 25mm or about 0.98 inches. So a PVC pipe with a 1 inch inside diameter will hold an eyepiece tube, but not the edge of the eyepiece. This is because the eyepiece has a slight overhang or flange which prevents it from falling into the tube.

Therefore, do not remove the eyepiece from its tube. Measure the outer diameter of the eyepiece with its flange. Use a caliper, or use another method. For example, you can place a ruler across the top of the eyepiece. Mine measures 28mm. You can also trace a circle on a piece of paper. You can even look around the house for a medicine bottle of the correct diameter. My medicine bottle was a perfect fit. You will need to bring your measurements to the hardware store.

Camera Adapter Tube Measurement

Now turn on your camera and let the lens pop out as in the photo below. Measure the outer diameter of the camera lens at its widest point. You must also measure the length that the lens extends from the camera body, as in the picture on the right. Important - put your camera setting on "macro" if that is what you will use when taking pictures. The macro setting may affect the length that the lens extends from the camera body.

If you are using a digital SLR, measure the widest part of the lens, where the zoom or focus ring is. Be sure to use the lens you will use for shooting macro shots. Also when measuring the distance from the camera body to the front of the lens, be sure to include the thickness of the UV filter, which you might have on the lens.

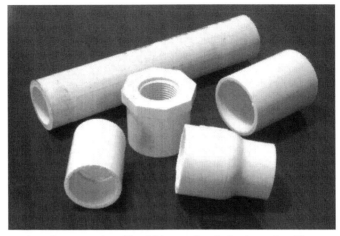

Your entire adapter project should be made of plastic, so it will not scratch the microscope or your camera. PVC plumbing parts are the best solution for this. Anything made of metal will scratch your microscope. Therefore, plan on a makeshift look, and spending very little money.

Selecting Your Components

1. A big piece of PVC for a camera cradle. Look for a large "end cap." Mine is a 4 inch inside diameter end cap designed to go over a large PVC pipe.

2. Eyepiece tube with inside diameter slightly bigger than your microscope eyepiece tube. I found a piece with an inside diameter of 28.2mm. It is designed to go on the outside of a 1 inch PVC pipe.

3. Camera adapter tube for camera connection to the eyepiece tube. You will have to get something that can go over your camera lens, and maybe another tube to adapt down to the above eyepiece tube. I skipped the additional tube because one fitting worked for me.

4. 3 plastic or nylon thumbscrew knobs.

5. PVC cement.

6. Plastic model putty. Wood putty is okay, but get the kind that will adhere to plastic.

Tools You Will Need

1. Eye protection (goggles or safety glasses).
2. Screw tap that matches the plastic thumbscrews.
3. A drill (electric or hand drill).
4. Hole saw drill, or an auger bit to drill a large hole.
5. Jig saw or band saw (a hand held jig saw is fine). A Dremel tool might work.
6. Vice.
7. File.
8. Sandpaper.
9. Vacuum cleaner.

Here are some suggested sizes and guidelines for the components:

For your camera cradle you need the biggest PVC platform you can find. Your camera will rest on top of it. I used an end cap with a four inch inside diameter. Depending upon the type of drill and saw you have, you can really buy anything you feel you can deal with. Remember that you will need to bore a large hole into it, and you will need to cut the sides off of it.

Below is the finished adapter, showing the end cap I used for a cradle. Black lines are where I can use a jig saw to trim off the sides of the cradle.

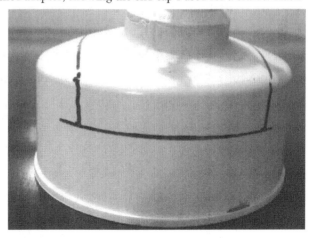

I was fortunate enough to find a camera adapter tube that was the correct size on both ends, so I could use the one component and no additional tubes. I attached that to the end cap, and I was basically done.

One end (about 1 inch inside diameter) fit over my microscope, and the other end held my camera lens perfectly. My camera for this set up was a Sony Cyber-shot 7.2 mega pixel point and shoot. Try out your parts in the store. Look for adapters and bushings.

You may not be so fortunate, and you might have to buy separate tubes and interconnect them. If that is the case, you will have to do quite a bit of testing one tube with another, as in the below photo.

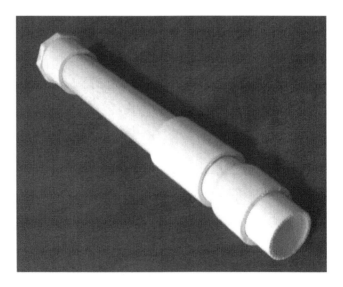

The goal is to have your eyepiece sit nicely inside one tube, and by nesting that into another tube or series, have the other end receive your camera lens. Then cut them all down as short as possible so they work, and are stable as well.

For your microscope eyepiece tube, the inner diameter of this tube must allow a free sliding, not too snug, fit over the eyepiece and the eyepiece tube. Buy a length of at least three inches in case you need to trim some.

For your camera adapter tube, as suggested, you may need to find possibly more than one piece to connect between camera lens and the eyepiece tube. So select your microscope eyepiece tube first, and then try to fit it into a pipe fitting that has an opening the size of your camera lens. There are many such PVC pipe adapter fittings already available, and no modifications may be necessary.

Bring your camera to the hardware store, and try it out with some plumbing

parts. Not to scare you, but you may have to make several trips to the store, returning the items that do not work. Save your receipts. Once you find the winning combination, you are done with this part of the task.

Next, find plastic or nylon thumb screws in the thin sliding drawer trays in the hardware aisle of the chain stores. In smaller hardware stores, just ask for them. The typical size is 8-32 and that works with a tap of the same size. On the previous page is a picture of the nylon thumb screw.

If you don't already have a screw tap, buy one to match the thumb screws. It's a slight investment; however you will be able to use this tool in the future. If you've never used a screw tap before, don't be intimidated. It was new to me as well. It's great fun to learn how to use a new tool, and it is  easier than you think. Mine is size 8-32. The screw tap comes in a set with a matching drill bit for making the hole.

Buy the smallest can of PVC cement you can find, and I recommend you buy the kind that does not require two cans (A & B) to work. PVC cement is messy, smelly, highly flammable, and toxic. And when you are done with it, you must bring it to your local town dump that accepts paint related materials like this. Do not throw it in the trash, because it is very bad for the environment and is a safety hazard. After you buy it, don't leave it in a hot car. Drive straight home with it and put it in your garage. Don't put it anywhere near a hot water heater.

Plastic model putty will adhere very well to PVC plastic, and this comes in small tubes which contain just the right amount for your project. You can find this in a big crafts store and most certainly in any hobby shop. You can also use wood putty, as long as it is plastic based and will adhere well to PVC plastic. You can read the back label or test some on a piece of PVC and see if it sticks. You will be using the putty to seal any cracks between the various components. Of course, when you are done you can sand and paint your masterpiece, and your adapter might look really great.

Putting It All Together

Drill a large hole, off center, in the bottom of your big camera cradle. Use an auger bit or a hole saw bit. You have to decide where to drill the hole based upon the position of the lens in your camera and how it will sit in or on the cradle. The idea is to be able to rest the camera in the cradle and have the lens hanging down through it.

The below picture shows where I drilled a hole in the end cap, inserted, then cemented the microscope adapter tube. The size of this hole will be exact size of the outer diameter of your camera adapter tube. The adapter tube must fit very snugly through the off-center end cap hole, so the PVC cement will work properly. You may have to drill the hole slightly small and then file and sand it outwards to the correct measurement of the adapter tube. When you drop your camera into the cradle it should line up  with the hole. You may also cut away the sides of the cradle with your jig saw or band saw to make your camera fit properly.

As explained above, my project was almost finished after cementing in just the one adapter tube. The underside side fit the microscope eyepiece and the top allowed a perfect fit for my Sony Cyber-shot camera. You may have to buy another camera just for your microscope photography.

You will have to work out the various tube combinations for your particular camera. Run tests by taping the parts together, and then, when you have it right, apply your PVC cement per the instructions on the can.

This is a trial and error process; however with patience and ingenuity, it will work. Use the putty to fill in any crack or gaps. Smooth it out and when the putty is dry, sand it. Let it dry in a well-ventilated place over night.

The last step is to drill small holes using the drill bit that goes with your particular screw tap. There will be three holes, evenly spaced around the bottom end of the microscope adapter tube. Drill these holes about ¾ inch from the bottom of the tube. The picture on the next page shows drilling the hole in the tube.

Now tap out the holes using the screw tap. You'll wind up with threaded holes like the one shown below.

Next, insert the plastic thumbscrews, screwing them in just enough to hold them in place, but not past the inside surface of the camera adapter tube.

The black lines show where I plan to cut the end cap sides to have more access to the camera controls. You can see above how all three thumbscrews will have to be unturned, so the insides are flush, before you slide the adapter over the microscope tube.

Now you are ready to use your adapter. You should be able to slide the entire assembly over the microscope and tighten the plastic thumbscrews to hold it in place. Tighten them evenly, by alternating one at a time, and do not over tighten.

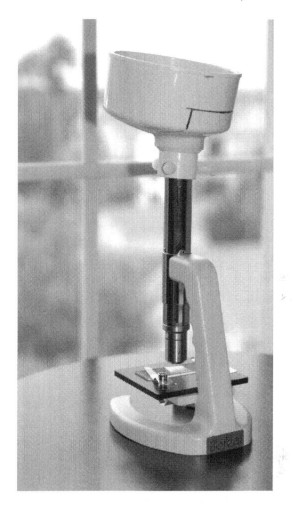

Turn on your camera, macro setting, and drop it into place. All should work perfectly. Many of the microscope pictures in this book were taken using the PVC adapter you see here.

CHAPTER 12

How to Make a Plant Press

Whenn you do research on tardigrades, it is important to describe and document the environment where tardigrades are found. Since most of your field samples will be from lichen on trees, tree identification becomes all important. Field manuals about trees are very helpful but, often, you will need to do more research to identify a particular tree species. Leaf samples are critical in this task. A good plant press is the answer.

Once you have a leaf press, you can use it for all sorts of projects, including crafts projects where you need to press things flat. You can press and make dried flowers and collect beautiful fall colored leaves for variety of projects. A plant press is so easy to make that, after you read this, it may be the first thing you do.

Below you will find an outline of the very simple process for making a plant press. We'll start with materials.

Selecting Your Components

- Long bolts (4)
- Washers (8)
- Thumbscrews (4)
- Wood planks (2) - each 12 x 16 inches or bigger
- Artist's watercolor paper, acid free (1 pad)

The diameter of the long bolts does not matter, as long as the washers fit nicely and the thumbscrews work on the ends. The length of the bolts will depend upon how thick you want to make your press.

Don't even bother measuring anything. Just go to the hardware store and look at their long bolts. Slide on the washers and screw on the thumbscrews. The total cost will be very reasonable. Because I built a thick press, I used very long bolts. I used heavy scientific journals for top and bottom padding. For your actual pressings, buy acid-free paper in an art supply store. I used a thick pad of archival quality watercolor paper. People who paint in watercolors use the perfect kind of paper which does not discolor with age, and which absorbs moisture. These pads are not too expensive either.

Pick out two nice planks of wood, and keep in mind, you'll probably have this press for the rest of your days, and maybe pass it on to someone. So get some nice wood that you can sand, stain, and varnish. The wood planks must be at least 12 inches by 16 inches or bigger. Sheets of standard paper have to fit inside. Pads of watercolor paper are usually 9 x 12 inches, so you have to allow an extra 2 to 3 inches on each side.

My daughter actually decorated my press with wood burning technique, which added to the beauty of the natural grain and dark knots in the wood. Wood burning is a nice hobby in itself, and you can make quite a beautiful press.

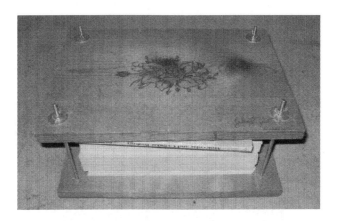

Building Your Plant Press

Drill four holes, one in each corner of your press. Be sure to space the holes far enough apart so that your paper, when inserted, will be clear of the holes. To do this, stack the boards and drill through two boards at a time. This way, your holes will be aligned perfectly. Then mark the edges of the planks in light pencil so you will be able to line up the two planks the same way later.

Under the bottom plank, place the four washers at the hole positions. Then slide the bolts, from bottom up, through the washer and then the wood.

Lay the top plank on top of the four bolts, sliding it down flush with the bottom plank.

Drop the remaining four washers over the bolts and screw on the thumb-screws about ½ inch.

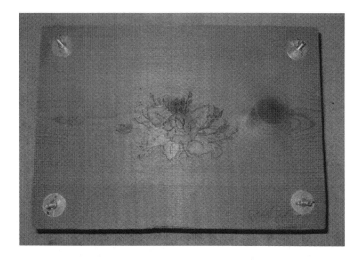

Now raise the top plank all the way up, as you insert your paper. You only need a few sheets of the good watercolor or archival paper in the center, and you can use old magazines and scrap sheets for top and bottom pressure.

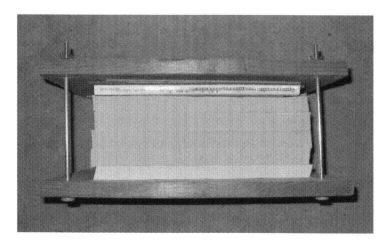

Now, enjoy taking flower, leaf and plant samples. After pressing for a couple of days, you can assemble the pressings in a book. You can even make photocopies, which come out beautifully in color, then scan those photocopies into PDF files for research or posterity.

Here is another picture of a pressing.

You can also gently remove the plants, and preserve them using clear hot-lamination material. This will allow you to study both sides of the plant, and even put it under a microscope (if you slice up the laminate into small microscope slide-sized pieces).

CHAPTER 13

Tardigrade Survey Scientific Paper

C harts, graphs, and tables do not always show up well in small printed paperbacks. I have left my original scientific paper intact here. You can also go to my website http://www.tardigrade.us for the original paper.

The Scientific Paper

TARDIGRADES OF NORTH AMERICA:
A PRELIMINARY SURVEY OF NEW JERSEY, U.S.A.

Michael W. Shaw, PO Box 742, Midlothian, VA 23113

William R. Miller, Department of Biology, Baker University, Baldwin City, KS 66006

ABSTRACT

Terrestrial tardigrade habitat samples were collected from fifty locations throughout the twenty-one counties of New Jersey, USA. Four species from four genera, within three families are reported. *Milnesium tardigradum, Macrobiotus hufelandi, Minibiotus intermedius, and Ramazzottius oberhaeuseri* are all considered cosmopolitan and found on most other continents. Tardigrades were found in all counties, in all types of habitats throughout the state.

Key Words: tardigrade, biodiversity, distribution

INTRODUCTION

Tardigrades (phylum *Tardigrada*) or water bears are microscopic aquatic animals found in the interstitial water of terrestrial habitats such as moss, lichen, leaf litter, and algae on tree bark. These habitats cycle between dry and moist, retaining moisture long enough to allow tardigrades to be active, grow and reproduce (Miller, 1997). When the habitat desiccates, the animal loses 97% of its body water, and shrivels into a structure called a "tun" until moisture returns. In this state, called "cryptobiosis," the water bear can survive extreme temperatures (-200C, +160C), high pressures (6,000 ATM), vacuum and excessive radiation (Kinchin, 1994). With their extreme survival capability, cryptobiotic tardigrades have recently become the first multi-celled animal to survive exposure to outer space (Jonsson et. al., 2007).

Tardigrades have five body segments, four pairs of legs, with claws on each leg. They have a dorsal brain and a ventral nervous system. They have a complex pharyngeal structure, complete digestive system, but lack a circulatory or respiratory system (Miller, 1997). Tardigrades have a single gonad and separate sexes but some genera are hermaphroditic. Water bears are ecdyzoia and shed their chitinous cuticle to grow; some lay their eggs in the cuticle as they shed, while others deposit their eggs free in the environment. They range in size from 0.3mm to about 1.2 mm (Kinchin, 1994). Despite more than 900 described taxa (Guidetti & Bertolani, 2005; Degma and Guidetti, 2007) there have been few systematic surveys of large (country sized) areas to document distributional patterns such as England (Morgan & King 1976) or Poland (Dastych, 1978). In the United States, few state sized surveys such as Illinois (Pugilla, 1964), California (Schuster & Gragrick, 1965), or western Montana (Miller, 2007) exist. Many states such as New Jersey have not been surveyed, thus the distributional patterns, environmental affinities, and habitat data is non-existent (McInnes, 1994; Miller 1997).

This project is a model for naturalist or citizen science exploration and demonstrates how true collaboration can lead to discovery and the expansion of knowledge. This is the first report of the existence of four species of the animals of the phylum *Tardigrada* in New Jersey. It expands the known range and distribution of each species, makes observations on habitat requirements, and adds to the known biodiversity of the region and state of New Jersey.

MATERIALS AND METHODS

Tardigrades were surveyed by sampling in all twenty-one New Jersey counties, between 2001 and 2009. Rural and urban sites ranging from parking lots, roadside trees, urban office complexes, nature preserves, interstate highway rest stops, and residential neighborhoods were selected (Table 1). A Magellan Global Positioning System (GPS) was used to fix sample locations. Most collection sites were photographed (68%).

COLLECT DATE	SPECIMEN No.	LATITUDE	LONGITUDE	ELEVATION (meters)	CITY/TOWN NAME	COUNTY	SAMPLE TYPE	SUBSTRATE
4/18/2001	8	40 DEG 50.455 N	074 DEG 28.541 W	195.12	Morris Plains	Morris	bark	Dawn Redwood *Metasequoia glyptostroboides*

Date	No.	Latitude	Longitude	Elevation	Locality	County	Substrate	Host
7/2/2002	9	40 DEG 54.244 N	074 DEG 49.382 W	401.10	Hackettstown	Warren	moss	rock
4/14/2003	1	40 DEG 38.856 N	074 DEG 16.142 W	28.82	Roselle	Union	moss	dirt
4/19/2003	2	40 DEG 38.856 N	074 DEG 16.142 W	28.82	Roselle	Union	moss	dirt
4/24/2003	3	40 DEG 20.356 N	074 DEG 28.958 W	60.94	Cranbury	Middlesex	moss	dirt
5/1/2003	4	40 DEG 50.455 N	074 DEG 28.541 W	195.12	Morris Plains	Morris	bark	Dawn Redwood *Metasequoia glyptostroboides*
5/9/2003	5	40 DEG 56.956 N	074 DEG 523 W	43.94	Fair Lawn	Bergen	moss	dirt
5/28/2003	6	40 DEG 1	74 DEG 20	35.43	Lakehurst	Ocean	bark	Silver Maple *Acer Saccharinum*
5/30/2003	7	40 DEG 37	74 DEG 34	195.12	Somerset	Somerset	bark, lichen	Norway Maple *Acer platanoides*

7/9/2003	7/9/2003	7/17/2003	7/17/2003	7/18/2003	7/23/2003	9/3/2003	9/26/2003
10	11	12	13	14	16	15	17
41 DEG 09.939 N	41 DEG 54.244 N	41 DEG 00.991 N	41 DEG 00.799 N	40 DEG 51.077 N	41 DEG 04.490 N	40 DEG 03.591 N	40 DEG 34.916 N
74 DEG 33.435 W	074 DEG 49.382 W	073 DEG 56.794 W	073 DEG 56.798 W	074 DEG 20.110 W	074 DEG 08.311 W	074 DEG 37.874	074 DEG 40.676 W
224.88	409.13	30.24	19.37	86.93	86.93	83.62	69.92
Hamburg	Hackettstown	Northvale	Northvale	Pinebrook	Ramsey	Cookstown	Bridgewater
Sussex	Warren	Bergen	Bergen	Essex	Bergen	Burlington	Somerset
moss	lichen	bark	bark	bark	bark	bark	bark
rock	rock	September Elm *Ulmus serotina*	Black Mulberry *Morus nigra*	Paper Birch *Betula papyrifera*	American Sycamore *Platanus occidentalis*	American Linden *Tilia americana*	Osage Orange (Bodark) *Maclura pomifera*

Date	No.	Latitude	Longitude	Value	Location	County	Substrate	Tree
10/27/2003	18	39 DEG 41.910 N	075 DEG 23.746 W	19.37	Auburn	Salem	bark, lichen	Pin Oak *Quercus palustris*
11/4/2003	19	40 DEG 26.951 N	074 DEG 40.261 W	65.20	Harlingen	Somerset	bark	Scarlet Oak *Quercus coccinea*
11/5/2003	20	40 DEG 52.178 N	074 DEG 26.690 W	184.25	Parsippany	Morris	bark, lichen	Hybrid Crab Apple *Malus hybrids*
11/12/2003	21	40 DEG 16.493 N	074 DEG 03.562 W	85.98	Eatontown	Monmouth	bark, lichen	Little leaf Linden *Tilia cordata*
11/12/2003	22	40 DEG 16.493 N	74 DEG 03.562 W	86.46	Eatontown	Monmouth	moss	dirt
11/14/2003	23	40 DEG 02.860 N	074 DEG 44.200 W	294.33	Newton	Sussex	moss	rock
11/18/2003	24	41 DEG 58.556 N	074 DEG 59.398 W	167.24	Sycamore Park	Warren	bark	American Sycamore *Platanus occidentalis*
11/19/2003	25	40 DEG 26.881 N	074 DEG 26.643 W	33.54	Brunswick	Middlesex	moss	dirt

Date	No.	Latitude	Longitude	Value	City	County	Substrate	Species
11/21/2003	26	40 DEG 40.432 N	074 DEG 26.454 W	130.87	Berkeley Heights	Union	moss	dirt, pavement
5/25/2004	27	40 DEG 11.328 N	074 DEG 09.838 W	34.49	Farmingdale	Monmouth	bark, lichen, tree moss	Sweet Gum Liquid Ambar styraciflua
5/25/2004	28	40 DEG 04.644 N	074 DEG 09.931 W	20.31	Lakewood	Ocean	bark, lichen	American Linden (Basswood) Tilia americana
5/25/2004	29	40 DEG 57.435 N	074 DEG 10.296 W	18.90	Toms River	Ocean	bark, lichen	Norway Maple Acer platanoides
6/3/2004	30	40 DEG 17.447 N	074 DEG 04.687 W	29.76	Ft. Monmouth	Monmouth	bark, lichen	Northern Red Oak Quercus rubra
6/3/2004	31	40 DEG 53.307 N	074 DEG 43.392 W	435.59	Budd Lake	Morris	moss	dirt
6/4/2004	32	41 DEG 00.150 N	074 DEG 14.638 W	197.48	Oakland	Bergen	bark, lichen	tree
6/4/2004	33	40 DEG 46.728 N	074 DEG 05.039 W	5.20	Secaucus	Hudson	Bark	Black Locust Robinia pseudoacacia

6/13/2004	8/5/2004	4/20/2005	4/20/2005	7/7/2005	7/28/2005	7/28/2005	9/1/2005
34	35	36	37	38	39	40	41
40 DEG 42.650 N	40 DEG 46.072 N	40 DEG 31.62 N	40 DEG 30.125 N	40 DEG 38.138 N	40 DEG 09.513 N	40 DEG 10.537 N	40 DEG 13.703 N
074 DEG 45.436 W	074 DEG 15.122W	075 DEG 03.560 W	074 DEG 51.295 W	074 DEG 54.732 W	074 DEG 25.704 W	074 DEG 35.218 W	074 DEG 37.183 W
263.62	102.05	61.89	87.40	103.94	77.01	39.68	31.65
Tewksbury	West Orange	Frenchtown	Flemington	Clinton	Jackson	Allentown	Robbinsville
Hunterdon	Essex	Hunterdon	Hunterdon	Hunterdon	Ocean	Monmouth	Mercer
bark, lichen	bark, lichen	moss	lichen	moss	lichen	moss	bark
Green Ash *Fraxinus pennsylvanica*	Scarlet Oak *Quercus coccinea*	American Sycamore *Platanus occidentalis*	Common Pear Tree *Pyrus Communis*	Black Locust *Robinia pseudoacacia*	Weeping Willow *Salix babylonica*	American Hornbeam	Sweet Gum *Liquid Ambar styraciflua*

Date	No.	Latitude	Longitude	Value	City	County	Substrate	Tree
1/11/2006	42	40 DEG 01.118 N	074 DEG 43.930 W	39.68	Unionville	Burlington	lichen	Maple
3/4/2006	43	39 DEG 29.999 N	074 DEG 31.813 W	35.43	Pomona	Atlantic	bark	White Oak *Quercus alba*
3/4/2006	45	39 DEG 10.893 N	074 DEG 43.400 W	1.89	Sea Isle City	Cape May	bark	American Holly *Ilex opaca*
3/5/2006	44	39 DEG 29.999 N	74 DEG 31.813 W	35.43	Pomona	Atlantic	moss	Sand
8/8/2007	46	39 DEG 39.604N	074 DEG 52.708W	65.67	Blue Anchor	Camden	moss	White Oak *Quercus alba*
8/8/2007	47	39 DEG 41.349 N	075 DEG 00.845W	42.99	Downer	Gloucester	moss	White Oak *Quercus alba*
8/20/2007	48	40 DEG 55.154N	074 DEG 13.755W	70.39	Totowa	Passaic	lichen	Red Maple *Acer rubrum*
8/20/2007	49	41 DEG 00.067N	074 DEG 16.516W	76.06	Wayne	Passaic	moss, lichen	White Oak Quercus alba

Date	No.	Latitude	Longitude	Value	City	County	Substrate	Object
9/6/2007	50	40 DEG 41.655 N	074 DEG 03.514W	0.94	Jersey City	Hudson	lichen	London Plane *Platanus × acerifolia*
9/25/2007	51	40 DEG 25.826 N	075 DEG 14.642 W	47.72	Bridgeton	Cumberland	lichen	White Oak *Quercus alba*
9/25/2007	52	39 DEG 25.736 N	075 DEG 14.824 W	40.63	Bridgeton	Cumberland	lichen	stone
9/25/2007	53	39 DEG 26.117 N	075 DEG 15.476 W	38.27	Hopewell	Cumberland	bark	Oak
9/25/2007	54	39 DEG 34.209 N	075 DEG 27.864 W	10.39	Salem	Salem	bark, lichen	American Sycamore *Platanus occidentalis*
3/7/2008	55	40 DEG 38.03 N	074 DEG 16.17 W	29.29	Linden	Union	bark, lichen	tree
6/25/2009	56	39 DEG 30.821 N	074 DEG 55.541 W	47.72	Buena	Atlantic	lichen	Brick
6/26/2009	57	40 DEG 32.287 N	074 DEG 17.673 W	72.28	Woodbridge	Middlesex	lichen	stone

6/26/2009	58	40 DEG 33.888 N	074 DEG 19.159 W	55.28	Woodbridge	Middlesex	lichen	stone

Table 1. New Jersey Collections Sites Sorted by Date

Samples were collected in paper "lunch bags" or small manila "coin envelopes." Moss samples were taken from dirt, rock, trees, stone (including asphalt), and sand substrata. Lichen, bark and algae was scraped from a tree, rock, brick, or stone substrate with a paring knife or a single edged razor blade. Scrapings went directly into a sample bag or envelope and were allowed to dry (Shaw, 2012). In some cases, 50mm deep plugs were collected by using a 62mm diameter soil corer, and the dirt substrate was retained (Figure No. 1).

Trees were photographed for height, crown shape, bark type, leaf characteristics, and branch pattern (Figure Nos. 4 through 7). When possible, leaf and seed pods were collected in season to aid in identification. No collections were made during December or February due to lack of foliage, however January was used as a collection month to represent winter for the survey (Table No. 2). Leaves were dried in a leaf press for one week using acid free paper, then color photocopied for detail preservation. Pressings were later photographed with centimeter ruler to aid botanical reference (Figure Nos. 5 and 6). Trees were identified to species with the keys of Brockman (1986), Coombes (1992), and Little (1980).

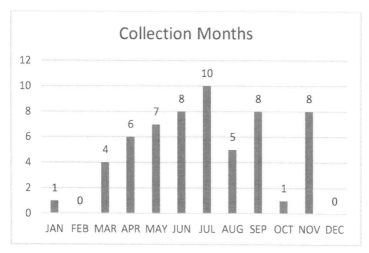

Table 2. Collection Months.

Samples of several grams each were placed in either plastic petri dishes or 2 oz. plastic cups and hydrated with spring water to a depth of 0.5 to 1.5cm. Each container was labeled with date, site name and specimen number. Samples were maintained at room temperature and examined for tardigrades after 6 to 48 hours, using a B&L Stereo-Zoom dissecting microscope, in a range between 20x and 30x magnification. Cross contamination was avoided when working with multiple samples by using a disposable pipette for each sample.

Individual tardigrades were transferred into a drop of media on a 25 x 75mm glass slide with an Irwin loop or eye-dropper (Shaw, 2012). Glass coverslips were applied, and live tardigrades were studied while on slides, observations recorded in notebooks. Permanent slides were prepared by replacing water with alcohol and/or solvents in stages, and finally with Hoyer's or PVA (Polyvinyl Alcohol) as final mounting media. Glass cover slips were applied and sealed with Cytoseal™ 60.

Tardigrades were examined and photographed with a Lomo Multiscope using bright-field, dark-field, incident lighting, phase contrast, polarized light, and Rheinberg illumination at various magnifications. Live tardigrades in vitro and fixed slide mounted tardigrades were photographed. Incident lighting shows the tardigrade as it appears most naturally in daylight, if one were able to observe it in nature with the naked eye (Figure No. 3). Cross polarization brightly illuminates only the stylets and certain minor internal structures (Figure No. 1). Phase contrast allows better determination of claw type when differentiating between species (Figure No. 8) Final species determinations presented here were made using a Differential Interference Contrast (DIC) microscope at various magnifications. Species identification was determined using the keys of Ramazzotti and Maucci (1983), and Nelson (2000).

Figure No. 1 Specimen No. 25. Background is site location with moss in foreground. Inserts (clockwise) show in vitro tardigrade close ups (brightfield lighting) of mouthparts, pharynx, macroplacoids and claws (indicating possibly *M. harmsworthi*); same in polarized light; central portion with claws in polarized light; moss close up showing where round core samples were removed.

RESULTS

New Jersey, a state in the United States of America, lies on the Atlantic Ocean to its east, and is bordered on the west by the Delaware River. The state at its center is positioned on the North American continent at approximately 40 degrees North Latitude. It has a temperate climate, with monthly average temperatures ranging from about 29.4 C (85 F) in summer to about 4.4 C (24 F) in the winter of most years. Spring and autumn are very mild. The mean rainfall for the last decade was recorded at about 124cm (49 in.), up about 5cm (2 in.) from the previous decade. New Jersey encompasses 22,610 square kilometers (8,729 sq mi.), of which 14% is water. Mean elevation of the state is 76.2 meters (250 ft.) above sea level.

New Jersey geography is varied, comprising four major types of landscape. The Atlantic Coastal Plains covers 3/5ths of the state, characterized by marshlands and meadows. The industrialized Piedmont region includes New Jersey's major rivers: The Hudson River, Passaic River, Ramapo River, and Raritan River. The Highlands have flat-topped rocky ridges with many lakes interspersed throughout the region. The Appalachian Ridge and Valley Region consists of mountains in the northwest, with valleys of shale and limestone.

Because of this wide diversity in geography and climate, New Jersey seems to provide an excellent place to survey for the presence of tardigrades when seeking to understand more about habitat and conditions that might contribute to ubiquity in this or any species.

Fifty-eight habitat site samples were collected from fifty locations in the twenty-one counties of New Jersey. Locations included forests, wetlands, urban areas, residential neighborhoods, river side and ocean side locales, business and commercial districts, at a variety of elevations (Table Nos. 1 and 3).

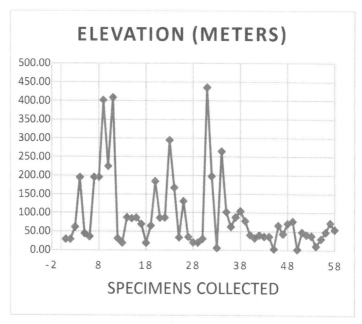

Table No. 3 Elevation Range. GPS data from site locations (Table No. 1) was entered into Garmin MapSource (c) software to generate a population grid. The grid points were transferred into Google Earth (Google, 2013). A county map image from Mapwatch.com (c) was then layered into Google Earth and aligned with the terrain and boundaries. This new composite map was then made opaque, and is presented in Figure No. 2.

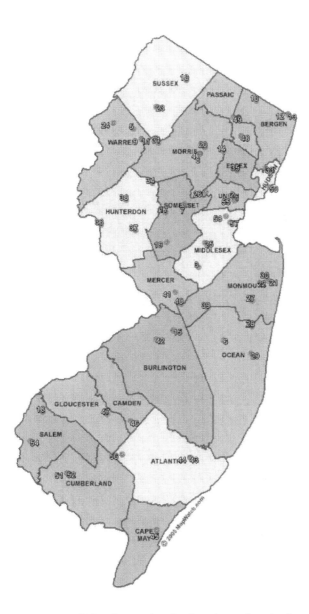

Figure 2. Map of New Jersey showing locations of study sites within their respective counties.

Collections took place over a period from 2001 through 2009 (Table Nos. 1 and 4). In four cases, samples were taken on the same day from different substrates at the same site; two locations were sampled on different days in the same month; two locations were sampled in different years, as shown in Table 1.

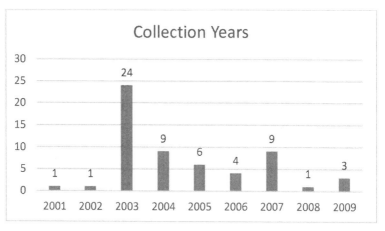

Table 4. Collection Years.

Average rehydration time for a dried sample was 4.9 days. Rehydration time to detect tardigrades ranged from 0.4 to 32 days (Table. 5), averaging 2.2 days. Specimen suspensions, 88 total, were discarded on an average of 10.3 days after no detection. Tardigrades were found in 50 of 58 (86%) site samples but in 7 cases tardigrades were lost before identification could be made, thus the number of positive identifications is only 43.

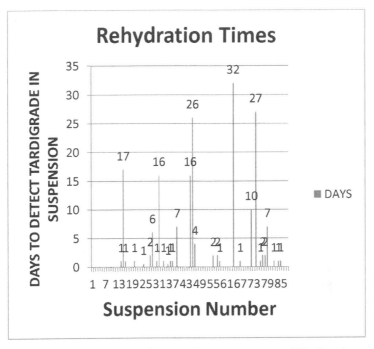

Table No. 5 Time from Start of Suspension to Detection of Tardigrade.

A total of 39 permanent slides were made, representing all New Jersey counties.

Although observations over the course of the study indicate five or more species*, we report thirty-three tardigrades and six eggs which represent four species: *Milnesium tardigradum, Macrobiotus hufelandi, Minibiotus intermedius, and Ramazzottius oberhaeuseri.*

Identification is based upon cuticle, claw type, lunules, buccal apparatus, and egg observations when eggs were present. Distribution data is presented in Table 6.

New Jersey Counties	Specimen No.	*Macrobiotus hufelandi*	*Minibiotus intermedius*	*Ramazzottius oberheauseri*	*Milnesium tardigradum*
Atlantic	43, 44, 56				X
Bergen	5, 2, 13, 16, 32	X			X
Burlington	15, 42	X		X	X
Camden	46	X			

County						
Cape May	45	X				
Cumberland	51, 52, 53					X
Essex	14, 35			X		X
Gloucester	47*	X				
Hudson	33, 50	X			X	
Hunterdon	34, 36, 37	X				X
Mercer	41*	X				
Middlesex	3, 25*, 57, 58	X		X	X	

Morris	4, 20, 31	X			
Monmouth	21,22 27,30,40	X	X		X
Ocean	6, 28, 29, 39				X
Passaic	48, 49*	X		X	
Salem	54		X		X
Somerset	7, 17, 19	X		X	
Sussex	10, 23	X			
Union	1, 2, 26, 55	X			X

Warren	11, 24					X

Table 6 . Tardigrades of New Jersey, Showing Distribution by County.

*In the case of specimen No. 41 (examined on a permanent slide), this may be *Macrobiotus harmsworthi* due to macroplacoid appearance and claw patterns. Eggs would be required for a more positive identification. Supporting that possibility is specimen No. 25, which in vitro appears to be *M. Harmsworthi* as well (Figure No. 1). Specimen No. 37 in vitro provided both an egg of *M.harmsworthi* as well as a tardigrade that could be *M. harmsworthi* (Fig. No. 3). Specimen No. 49 provided an egg appearing to be *Macrobiotus areolatus* along with a tardigrade that appears to be *M. areolatus* (Figure No. 4).

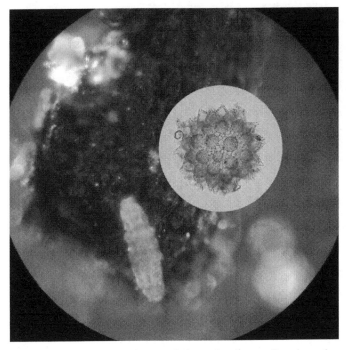

Figure 3. Specimen No. 37. Background shows in vitro tardigrade under incident lighting. Insert shows Brightfield photo of in vitro egg of *M. Harmsworthi* found with this specimen.

Figure No. 4 Specimen No. 49. Background is site location. Inserts (clockwise) show leaf close up; Brightfield photo of in vitro tardigrade close ups of mouthparts, pharynx and macroplacoids; claws; body (all indicating *Macrobiotus areolatus*); tree trunk detail; egg.

DISCUSSION

The only record of a tardigrade from New Jersey is a fossil, *Milnesium swolenski* (Bertolani and Grimaldi, 2000) found in Turonian amber near Sayerville in 1998. The specimen is estimated to be more than 90 million years old and documents that tardigrades in their existing form have been present in New Jersey since Upper Cretaceous (Turonian) (Bertolani and Grimaldi, 2000).

Although tardigrades were found in all counties within New Jersey, diversity was limited. This distinct lack of diversity across many types of rural, urban, coastal and central sites, and upon a variety of substrata, does not allow any major conclusions regarding how tardigrade species are distributed or dispersed. Blaxter, et. al. (2003) suggested that tardigrades in the size range under 1mm may be among the taxa that are ubiquitous and lack biogeography, as supported by Finlay (2002).

Substrata in this study were divided into five categories:

- Brick or Stone, man-made, from which a moss, lichen or algae sample was taken
- Dirt, from which a moss sample was taken
- Rock, natural, from which a moss, lichen or algae sample was taken
- Sand, from which a moss sample was taken
- Tree, from which a moss, lichen or algae sample was taken

Tardigrades were found to be present on all five substrata (Table No. 6).

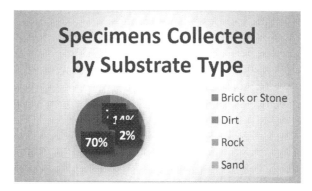

Table No. 6 Percentage of specimens collected by substrate type.

A further observation can be made about tardigrade distribution related to tree bark substrate. Trees with thin or smooth bark such as the Paper Birch (Betula papyrifera), and the Black Locust (Robinia pseudoacacia) tended not to support tardigrade presence despite multiple samplings. By contrast, the deeply furrowed bark of the Dawn Redwood (Metasequoia glyptostroboides), yielded tardigrade rehydration two years after specimen collection.

Figure No. 5 Specimen No. 20. Background is site location. Inserts (clockwise) show leaf pressing; fruit detail; Brightfield photo of vitro tardigrade mouthparts; posterior claws; pharynx and macroplacoids; body (indicating Macrobiotus hufelandi); tree trunk detail.

Figure No. 6 Specimen No. 39. Background is site location. Inserts (clockwise) show leaf pressing; Brightfield photo of in vitro tardigrade body; posterior claws; mouthparts, pharynx and macroplacoids (indicating *Minibiotus intermedius*); tree trunk detail.

Figure No. 7 Specimen No. 42. Background is site location. Inserts (clockwise) show Brightfield photo of permanent slide mounted tardigrade claws, pharynx and macroplacoids; full body showing distinctive brown color markings (indicating Ramazzottius oberhaeuseri); tree trunk detail.

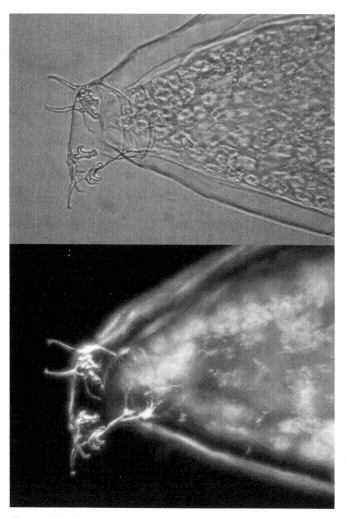

Figure No. 8 Specimen No. 35. Phase Contrast photo (above) of posterior claws (indicating *Milnesium tardigradum*) and Darkfield photo (below) of same.

LITERATURE CITED

Bertolani, R. and Grimaldi, D. 2000. A New Eutardigrade (Tardigrada: Milnesiidae) in Amber from the Upper Cretaceous (Turonian) of New Jersey. IN: Studies on Fossils in Amber, with Particular Reference to Cretacous of New Jersey, (ed. Grimaldi, D.). Backhuys Publishers, Leiden, Netherlands, p. 103-110.

Blaxer, M.., Elsworth, B., & Daub, J. 2003. DNA Taxonomy of a Neglected animal Phylum: an Unexpected Diversity of Tardigrades. Proceedings of the Royal Society of London, Biology Letters.

Brockman, F.C. 1986. Trees of North America, Racine, WI, Western Publishing Co., pp. 280.

Coombes, A.J. 1992. Trees, New York, Dorling Kindersley, pp. 320.

Dastych, H. 1978. The Tardigrada of Poland. Monografie Fauny Polski, Polska Akademia Nauk Zaklad Zoologii Systematycznej Doswiadczalnej. 16:1-255.

Degma, P. & Guidetti, R. 2007. Notes on the current checklist of Tardigrada. Zootaxa, 1579:41-53.

Finlay, B.J. 2002. Global dispersal of free-living microbial eukaryote species. Science 296:10611063.

Guidetti, R. & Bertolani, R. 2005. Tardigrade Taxonomy: An Updated Checklist of the Taxa and a List of Characters for their Identification. Zootaxa, 845: 1-46.

Jonsson, K. I., Rabbow, E., Schill, R.O.,Harms-Ringdahl, M., Rettberg, P. 2007. Tardigrades survive exposure to space in low Earth orbit. Current Biology, 18(17):729-732.

Kinchin, I.M. 1994. The Biology of Tardigrades, London, Portland Press, London, pp. 186.

Little, E. 1980. National Audubon Society Field Guide to North American Trees, Eastern Region. New York, Alfred A. Knopf, Inc. pp. 714.

McInnes, S.J. 1994. Zoogeographic distribution of terrestrial/freshwater tardigrades from current literature. Journal of Natural History, 28:257-352.

Miller, W.R. 1997. Tardigrades, Bears of the Moss. The Kansas School Naturalist, 43(3):1-16.

Miller, W.R. 2007. Tardigrades of North America: Western Montana. Intermountain Journal of Sciences, 12(3-4):27-38.

Morgan, C.I. & King, P.E. 1976. British Tardigrades. Tardigrada Keys and Notes for the Identification of the Species. Synopses of the British Fauna (New Series), London: Academic/Linnean Society of London, 9:1-133.

Nelson, D.R. & Marley, N.J. 2000. The Biology and Ecology of Lotic Tardigrada. Freshwater Biology, 44:93-108.

Pugilla, C.R. 1964. Some Tardigrades from Illinois. Transactions of the American Microscopical Society, 83(3):300-311.

Ramazzotti, G. & Maucci, W. 1983. Il Phylum Tardigrada. Memorie dell'Istituto Italiano di Idrobiologia 41:1-1011.

Schuster, R.O. & Gragrick, A.A. 1965. Tardigrada from Western North America: With Emphasis on the Fauna of California. University of California Publications in Zoology, 76:1-67.

Shaw, M.W. 2012, How To Find Tardigrades, Amazon Kindle, http://www.amazon.com/dp/B009D7VHXK, [original PDF version available upon request from the author] pp 21-24, 42-52., Retitled and revised 2014: Kids & Teachers Tardigrade Science Project Book.

.

Additional Information

Thank you to my wife Donna who contributed some of the photos, and to my two daughters who participated in my scientific journey.

And a big thank you to Erin, to whom I dedicate this book, for teaching me that millions of people just might be interested in what I have to say.

Other books of similar interest by Michael W. Shaw:

Your Microscope Hobby - How to Make Multi-Colored Filters

Kids & Teachers Tardigrade Quiz and Fact Book

Word Nerd – Things Way Up High Quiz & Fact Book

Websites:

www.tardigrade.us www.mikeshawtoday.com

www.amazon.com/author/mikeshaw247

http://astore.amazon.com/mikesmicroscopestore-20

ABOUT THE AUTHOR

Michael Shaw took a serious interest in science when his children were in middle school, and he helped them with their science projects. This led to a personal passion to make a contribution, and thus began a population survey of the little known creature (at the time), the tardigrade. The survey, covering the state of New Jersey, and publication of a scientific paper took about 10 years from start to finish.

This resulted in a great deal of how-to knowledge and a viral internet video popularizing the tardigrade as the "first animal to survive in space." He has starred in the PBS video on YouTube, "Songs For Unusual Creatures," and has appeared on television in Brasil. Having popularized this unusual creature, resulting in an exhibit at the American Museum of Natural History, he is known by the public as The Space Bear Hunter. He was recently featured in a science article in the New York Times.

Mr. Shaw has written several books to assist teachers and their young scientists in similar projects. Continuing to write, he is currently working on a novel and several other inspirational books.

Made in the USA
San Bernardino, CA
27 November 2018